MARE Publication Series

Volume 8

The MARE Publication Series is an initiative of the Centre for Maritime Research (MARE). MARE is an interdisciplinary social-science network devoted to studying the use and management of marine resources. It is based jointly at the University of Amsterdam and Wageningen University (www.marecentre.nl).

The MARE Publication Series addresses topics of contemporary relevance in the wide field of 'people and the sea'. It has a global scope and includes contributions from a wide range of social science disciplines as well as from applied sciences.

Topics range from fisheries, to integrated management, coastal tourism, and environmental conservation. The series was previously hosted by Amsterdam University Press and joined Springer in 2011. The MARE Publication Series is complemented by the Journal of Maritime Studies (MAST) and the biennial People and the Sea Conferences in Amsterdam.

Editors
Svein Jentoft
University of Tromsø, Norway
Svein.jentoft@uit.no

J. Maarten Bavinck
University of Amsterdam, The Netherlands
J.M.Bavinck@uva.nl

More information about this series at http://www.springer.com/series/10413

Knut Bjørn Lindkvist • Torbjørn Trondsen
Editors

Nordic-Iberian Cod Value Chains

Explaining Salted Fish Trade Patterns

Springer

Editors
Knut Bjørn Lindkvist
Department of Geography
University of Bergen
Bergen
Norway

Torbjørn Trondsen
University of Tromsø
Tromsø
Norway

ISSN 2212-6260 ISSN 2212-6279 (electronic)
MARE Publication Series
ISBN 978-3-319-16404-5 ISBN 978-3-319-16405-2 (eBook)
DOI 10.1007/978-3-319-16405-2

Library of Congress Control Number: 2015938766

Springer is a brand of Springer International Publishing

Springer Cham Heidelberg New York Dordrecht London

Printed on acid-free paper

Springer is part of Springer Science+Business Media (www.springer.com)

Preface

This book describes the changes and the driving forces in the trading pattern of codfish (Bacalao) between the Nordic and the Iberian countries. This trade has a long historic origin based on the food demand originated from the catholic people's preference for seafood in the lent season and the availability of big cod resources in the North Atlantic.

The bacalao trade has been an important contribution for the human existence and heritage over centuries in the fishing communities in Norway, Iceland, Faroe Island and Canada as well as for the southern Europe cuisine tradition. Many of the authors have a background in this culture; others have developed their cultural understanding through their research.

The book is a result of long term academic cooperation between the Universities of Bergen, Tromsø, Salamanca, Alicante and Santiago de Compostela and the former University College of Finnmark (now part of University of Tromsø, The Arctic University of Norway).

The main funding for the research project came from the Norwegian Research Council (Food Programme). The participation and partial project financing from the Export Council of Norway and the Norwegian Fishermen's Sales Organisation were also important for the research-based process. Sparebank 1 Nord-Norge's endowment fund gave also an important financial contribution for the Salted Fish Academy course in Galicia, Spain through the Seafood Centre at Vardo High School, Nordkapp Maritime College and High School and the then University College of Finnmark.

Particularly important project supporters were the manufacturers of bacalao Cape Fish Group from Honningsvåg, with its director Bjørn Ronald Olsen and his production manager Bernt Nicolaisen, as well as the firm Norfra from Tromsøand its director Steinar Eliassen and the leader of its Nordvågen facility, Odd Magne Nylund.

In Spain the project was considerably influenced by the organization ANFABASA-Asociación Nacional de Fabricantes de Bacalao y Salazones and their member manufacturers.

The project was managed by the Department of Geography at the University of Bergen. Siren Juliussen was project assistant in the project's first section. Research technician, Kjell Helge Sjøstrøm contributed with many of the book's maps and figures. The head of department, Svein Olaf Dahl, the head of administration, Gro

Aase and the economic consultant Marianne Soltveit have been important supporters of the project. The editors and project managers would like to thank all the institutions and individuals mentioned, for their positive participation and cooperation.

Finally, let us also mention two key names in the effort to transform research reports to scientific publications.. Professor Roger Hayter from Simon Fraser University in Canada was very important as a scientific advisor in the writing process. Dr. Heidi Bjønnes Larsen performed a key part of the editorial work on the first editorial phase. We thank them for their contribution.

Finally, we would like to express our thanks to our co-authors!

Let's just finally remind you that all scientific conclusions and formulations are solely the responsibility of the individual authors!

Bergen and Tromsø Knut Bjørn Lindkvist and Torbjørn Trondsen
20.08.2014

Contents

Contributors

Ana Espinosa Seguí University of Alicante, Alicante, Spain

Heidi Bjønnes Larsen Department of Geography, University of Bergen, Bergen, Norway

University of Bergen, Bergen, Norway

Manuel González-López Department of Applied Economics, Universidade de Santiago de Compostela ICEDE Research Group, Santiago de Compostela, Spain

Knut Bjørn Lindkvist University of Bergen, Bergen, Norway

Finnmark University College, Finmark, Norway

Department of Geography, University of Bergen, Bergen, Norway

Inmaculada Martínez Alba University of Seville, Seville, Spain

Øystein Myrland School of Business and Economics, University of Tromso, Tromso, Norway

Einar Richter-Hanssen University of Bergen, Bergen, Norway

José Luis Sánchez-Hernández University of Salamanca, Salamanca, Spain

Department of Geography, University of Salamanca, Salamanca, Spain

Torbjørn Schei Finnmark University College, Alta, Norway

Alexandre Trigo Department of Applied Economics, Universidade de Santiago de Compostela ICEDE Research Group, Santiago de Compostela, Spain

Torbjørn Trondsen University of Bergen, Bergen, Norway

University of Tromsø, Tromsø, Norway

Norwegian College of Fishery Science, University of Tromsø, Tromsø, Norway

Department of Geography, University of Bergen, Bergen, Norway

Sebastián Villasante Department of Applied Economics, Universidade de Santiago de Compostela ICEDE Research Group, Santiago de Compostela, Spain

Jinghua Xie School of Business and Economics, University of Tromso, Tromso, Norway

Chapter 1
Introduction

Knut Bjørn Lindkvist and Torbjørn Trondsen

Abstract The introductory chapter provides a brief overview of the purpose of this book about salted fish trade between Norway and Spain. The changing demand over time for fish products in Spain has impacts on both production systems and market levels in all countries participating in this specific value chain. The chapter thus introduces a first presentation of the historical-thematic basis of current trading with salted fish between Norway and Spain, and ultimately shows, through a brief presentation of the book's individual chapters, how the task of dealing with changing market dynamics is being solved.

Keywords Cod in the history · Research methods applied · Individual chapters

1.1 The History, the Trade and the Research

The fish trade has been important in Europe for centuries. After having wiped out fresh water fish in the eleventh century A.D. (Roberts 1997), Europeans took to the sea to fish, both for consumption and for trade (Kurlansky 1998, 2001). Dried cod enabled the Vikings to fish the northern hemisphere for centuries, marketing it to other parts of Europe (Kurlansky 1998). The Basques also became chief cod fishers in the medieval period. When John Cabot visited Newfoundland in 1497 and supposedly "discovered" this part of the world, he noticed 1000 Spanish cod fishing vessels lying in port. The development of salting techniques had made it possible for Basque fishers to explore these rich banks off Newfoundland to satisfy the European market. Although they had been frequenting the region for centuries, in order to safeguard their productive fishing grounds off Greenland, they had kept the location

K. B. Lindkvist (✉) · T. Trondsen
University of Bergen, Bergen, Norway
e-mail: Knut.Lindkvist@geog.uib.no

K. B. Lindkvist
Finnmark University College, Finmark, Norway

T. Trondsen
University of Tromsø, Tromsø, Norway

K. B. Lindkvist, T. Trondsen (eds.), *Nordic-Iberian Cod Value Chains*,
MARE Publication Series 8, DOI 10.1007/978-3-319-16405-2_1

and the newly discovered lands a secret (Kurlansky 1998). By 1550, 60 % of all fish eaten in Europe was cod and the salted fish trade between Norway and Spain has been particularly important since the sixteenth century (Richter-Hanssen Chap. 3). The Norwegian market share of salted fish products in Spain has declined from about 13,000 t in the early 1990s to 5000–7000 t in the late 2000s, while Iceland increasingly has taken over the market shares.

Salted cod is still very important for Norwegian fisheries and accounted for 48 % of the total export values of cod in 2012. The main market was Portugal which accounted for 39,000 t in 2012. Of all salted fish exports in 2012, 40 % went to Portugal and Spain (Norwegian Seafood Council 2013). Salted fish is an important product on the Iberian Peninsula. In both Spain and Portugal, salted cod (or *Bacalao/Bacalhau*) is considered as a distinct food category, different from other kinds of fish. It is sold in special shops and has dedicated stalls in the marketplace and in supermarkets. *Bacalao* has a unique set of recipes and plays a central role on religious occasions such as Christmas and Lent. Despite the significance of *Bacalao*, consumption of the traditional salted fish product on the Iberian Peninsula has decreased with the development of new products.

The global fish trade itself is not a new phenomenon. This book's main purpose is to give some insight into the bacalao trade from Norway to Spain and the changing dynamics in the embedded value chains over time. We are especially interested in identifying possibilities for the Norwegian salted fish industry to recapture this market.

Various chapters explain here the changing consumer demands in Spain and the inability of Norwegian processors to respond to these changes on both the product and institutional level. The book follows various lines of the cod trade and in different regions: Norway, Iceland and Spain. Within Spain, the book highlights the regional differences in consumer preferences that are influencing their final trade partners. Specifically, it offers several suggestions to Norway and to Norwegian processors to enhance the market share of Norwegian cod in Spain.

The authors represent a research project consisting of ten researchers and a number of accompanying students, all of whom collaborated on issues and discussions of findings. Networking and mutual quality control characterize the research method applied in the findings presented in this book. A network method is the one concept that best describes the mutual quality control and knowledge transfer that has taken place, in addition to the specific scientific methods applied in each chapter. In addition to the mutual exchange of knowledge and experience, the partners in the project and invited business leaders in focused industries contributed with their knowledge, criticism and development of research problems. An advantage of this method was that we developed a common understanding of key concepts, where the concepts of conventions and value chains were the most important.

Between 2008 and 2011, five intensive workshops were held with most of the researchers and partners, and two major field courses were arranged in Barcelona in 2009 and in Santiago de Compostela in 2010. External researchers, in addition to the members of the scientific group, have evaluated the draft of each chapter. Field courses which are referred to have been arranged by the former Finnmark University

College and contributed to the production of scientific reports that have helped to test some of the hypotheses in the project.

1.2 The Structure of the Book

This book is organized as follows:

Chapter 2 outlines the theoretical perspectives and the main concepts and models applied in the study, each with its basis in institutional economics and economic geography.

The empirical Part I describes the historical background, technological changes and institutional contexts in the Spanish market. This sets the scene and shows how Norway has prioritized quantity over quality and ignored the value of, and need for, education and knowledge in a traditional resource-based industry.

Chapter 3 deals mainly with a finding that path-dependent socio-economic processes in regions are a result of the specific history of those places. Production along the trajectory seen in both the Spanish market today, and the fate of the regions producing salted fish in Norway, began 350 years ago. The chapter describes the history and institutions that have formed the relationship between Norwegian salted fish producers and the Spanish market since their first contact in the late sixteenth century.

Part II focuses on the end of the value chains downstreams and the final market in Spain.

Chapter 4 investigates the food value chain. The main finding is that the value chains that are most likely to succeed are those able to agree about an efficient arrangement of the production system, and have a coherent understanding of the attributes that consumers expect to find embedded in its core foodstuffs.

Chapter 5 presents regional profiles of the heterogenic Spanish market. The main finding is that the Spanish salted fish market is a mosaic of several smaller regionally rooted markets rather than representing one market. The chapter presents its regional consumer conventions and quality preferences.

Chapter 6 mainly finds that the consumer conventions are moving towards a more healthy diet and towards clean food and sustainable fisheries. It investigates the consequences that changes in conventions can have for a value chain, focusing particularly on green conventions.

Chapter 7 investigates the connection between the Norwegian production system and the Spanish market demand. The main finding is that lack of success in the Spanish market is related to the Norwegian institutions and production and governance structures that seem designed for other purposes than satisfying the preferences in the Spanish consumer market.

Chapter 8 considers technological changes in the value chain. The chapter's main finding is how market and institutional change stimulated innovation by reducing uncertainties in the Spanish cod fishing industry. Nevertheless, innovation also implies the alteration of existing routines, habits and even legal frameworks. This chapter shows situations where an industry needs to break the existing institutional arrangements (formed by conventions like rules, norms etc.).

Chapter 9 analyses and compares two different development paths of value chains for codfish, respectively connecting Iceland and Norway to the Spanish market. The main finding is the identification of how path dependency in the Norwegian seafood value chain strengthens traditional production practices and limits market orientation. The value chain is however constituted of complex industrial structures, industrial business conventions which decide their value adding performance independent of the demand and product wishes in the Spanish market.

Chapter 10 illustrates the connection between production conventions, performance and consumer conventions. The main findings, demonstrated by econometric analyses, were that a significant increase in the consumption of frozen light salted fillets, at the expense of traditional products, has driven the restructuring of the Spanish saltfish market.

Chapter 11 tests the existence of Icelandic market power in Spain through descriptive data analysis and econometric modelling. The chapter's main finding is identification of the connection between Iceland's market power and consumer preferences in Spain and between institutions and values.

Part III focuses on the upper part of the value chains upstream of the resources and processing.

Chapter 12 focuses on educational vocational training (EVT). The main finding is that in Norway there has been a devastating turnaround for EVT in the fishing industry, where it has gone from a leading, vibrant role to one that is in decline. Knowledge is an important production factor and learning is a central practice in traditional resource-based industries. The chapter investigates the political economy and practice of the fishing industry, and the development of an educational infrastructure in northern Norway. The chapter claims that the currently matching conventions between the government and academia explain an academically oriented educational infrastructure that is not necessarily supporting the need for more resource-based market-oriented industrial knowledge development.

Chapter 13 investigates the structures that constrain the opportunities of seafood suppliers for satisfying consumer demand. The main finding is that the ability of managers to control transactions between input suppliers and paying customers are most important for performance in the value chain. This control is limited by six types of common business management convention: Commercial, Industrial, Domestic, Public, Ecological, and Civic (CIDPEC).

Finally, the concluding Chap. 14 sums up four decisive changes in the deep structure drivers and barriers in the Spanish market for salted fish. (1) The declining demand for traditional salted fish products. (2) The composition of the demand in the market. (3) The geographical patterns of the dominant consumer conventions. (4) The lack of market oriented adaption of Norwegian-Spanish value chain conventions. Finaly, the chapter discusses possible strategies for the producers to create interactive convention platforms between production knowledge and market knowledge in order to develop competitive, sustainable and profitable marketing strategies.

References

Kurlansky M (1998) Cod: a biography of the fish that changed the world. Penguin Books, London. ISBN-10: 0140275010

Kurlansky M (2001) The Basque history of the world: the story of a nation. Penguin Books, London. ISBN 0-8027-1349-1

Roberts CM (1997) Ecological advice for the global fisheries crisis. Trends Ecol Evol 12:35–38

Chapter 2
Analysing Food Chain Development: A Theoretical Framework

Heidi Bjønnes Larsen, Knut Bjørn Lindkvist, José Luis Sánchez-Hernández and Torbjørn Trondsen

Abstract This chapter presents a theoretical- analytical basis for investigation and discussion of the Norwegian-Spanish salted fish trade. The focus is on convention theory and value chain theory as analytical tools for this specific trade on salted fish. Special emphasis is put on Performance Drivers in the value chain and on specific contextual convention models for this salted fish trade through the value chain in Norway to Spain trade relations.

Keywords Sea food · Economic geography · Theory of conventions · Food chain development · Markets · Performance drivers

2.1 Introduction

The seafood trade from Norway to Spain is in this book analysed in a perspective framed by an economic geography that includes material, ecological, and socio-economic aspects of human life. This chapter will develop specific analytical concepts and models used in this book by focusing on drivers and conventions for value chain performance. Section 2 summarizes the role of economic geography as a discipline. Section 3 introduces the value chain concept where economic behaviour is explored in the context of a Structure-Convention-Performance (SCP) value chain model in which the term convention replaces the term conduct in the standard SCP model (Porter 1980; Barney 1996). Section 4 dicusses the use of broader sociological conventions concepts, related to different kinds of institutions, networks, values, and qualities, which are central for a deeper understanding of the

T. Trondsen (✉) · H. B. Larsen · K. B. Lindkvist
University of Bergen, Bergen, Norway
e-mail: torbjorn.trondsen@uit.no

T. Trondsen
University of Tromsø, Tromsø, Norway

K. B. Lindkvist
Finnmark University College, Finmark, Norway

J. L. Sánchez-Hernández
University of Salamanca, Salamanca, Spain

K. B. Lindkvist, T. Trondsen (eds.), *Nordic-Iberian Cod Value Chains*,
MARE Publication Series 8, DOI 10.1007/978-3-319-16405-2_2

situation of Norwegian salted cod in Spain and for any future attempt to recapture this market. Section 5 elaborates further on the interface between the value chain and the spatial and contextual properties of the conventions commonly used in food trade literature. The remaining chapters are based on this theoretical framework. Our focus is on discussion of empirical issues regarding history, prices, exports, companies, consumption, lock-in, education, and coordination.

2.2 Economic Geography as an Analytical Framework

Production, trade, location and growth have traditionally been the main concerns in the discipline of economic geography until the late 1960s. The inclusion of political economy in the 1970s unveiled the role of power in the production and reproduction of the structural inequalities intrinsic to the capitalist system (Barnes 2000, 2001; Scott 2000; Sánchez 2003; Foster et al. 2007; Barnes and Farish 2006). During the 1980s, the 'regulatory turn' emphasized the interplay between economy, society, polity and technology in an alleged transition from Fordism to post-Fordism. From the 1990s, the 'institutional turn' has attempted to explain the geographical unevenness of innovative capabilities and the subsequent variety of (national, regional, and local) development trajectories with reference to the so-called 'institutional thickness', commonly defined as the density and interlocking of public and private organizations that supply material and non-material resources to economic actors. At the same time, the 'cultural turn' addresses the role of culture and difference in the economic process and in the performance of economic actors. Even the more recent 'relational' and 'evolutionary' turns underline the 'softer' issues of organization, interaction, innovation, variety, and evolution as horizontal perspectives that may inform most research in economic geography.

Such quick theoretical and methodological renovation has allowed for inclusion of a number of new concerns in the discipline: poverty, (under)development and centre-periphery relations, technology, regulation, distribution, governance, labour, firms, policy, innovation, learning, sustainability, consumption, retailing, exclusion, or education actually constitute sub-fields of their own. Economic geography still pays attention to its core topics, of course, but geographers have also sought explanations beyond the traditional understanding of the production-distribution-consumption framework, thus acknowledging that the economy is an integrated part of the society influenced by social trends, facts and features.

The increasing attention paid by to local resources, competences, and values as factors that shape the realm of production has also required focus on the influence of power, institutions, networks and culture. The endowment of different types of local resources, namely whether inputs are unsuitable for standardization, is thus better accounted for by these broader theoretical approaches. In this vein, the study of territory-specific case studies may contribute to a deeper understanding of local contingencies stemming from the interplay between economy, society, culture and places.

Therefore, economic geographers all over the world are actually dealing with new terms, concepts, and theories that come from related disciplines such as sociology, anthropology, or cultural studies, let alone economics as the most common provider of vocabulary. This book is a good example of such a multi-disciplinary path: the Norwegian salted fish value chain and its performance in the Spanish food market is analysed in a broader social and institutional context, which both fosters and hampers adaptation, innovation and change. The main applied analytical frameworks to understand and explain the Norwegian and Spanish constraints and opportunities for economic development within the salted fish business relies on theories of industrial and value chain performance (Porter 1980, 1990), theories of convention (Storper and Salais 1997) and the theories of economics of justification (Boltanski and Thévenot 1991/2006).

2.3 Performance Drivers in Value Chains

Value chains are the main structure in the core of diversified and specialized economic activities linking natural resources, production and distribution systems to market needs (Porter 1985).

The traditional definition of a value chain is as a "complex set of interrelated activities required to produce a good or service and distribute it to consumers" (Hayter and Patchell 2011, p. 15). Independent, profit seeking buying and selling entrepreneurs and organisations are the driving force in each value chain wheel between the point of primary production and the final consumer markets. Each wheel in the value chain may also take advantage of resouces in specialized production regions. For example, fishing and primary processing are placed close to fishing grounds, while retailing takes place close to consumers in big cities. Each firm in the chain may be categorized in strategic groups, like fishing trawlers or wholesalers, which are following the same strategies and are exposed to similar competition with respect to limited production factors and customers (Porter 1985). Each production stage in the value chain is thus a competition arena directed by some common governing conventions and strategies as illustrated in Fig. 2.1.

Figure 2.1 illustrates that value chain structures are binding supplying and consuming regions together. The driving fuel in the value chain wheels is the exchange of two contra-flowing value streams, where products and services are flowing downstream from harvest to the consumers met by a counter upstreaming purchasing power from consumers and other financial suppliers.

The methods and contents of transaction decisions in all stages of a value chain relies however on established conventions defined as "...practices, routines, agreements, and their associated informal and institutional forms which bind acts together through mutual expectations" (Salais and Storper 1992, p. 174). Conventions can be both informal and formal. Informal as commonly accepted rules of behaviour that form the basis of shared cognitive frameworks that allow for coordination among economic actors (Storper and Salais 1997). The formal rules are 'more visibly' codified, written down, and supported by the state (Storper 1997, p. 72).

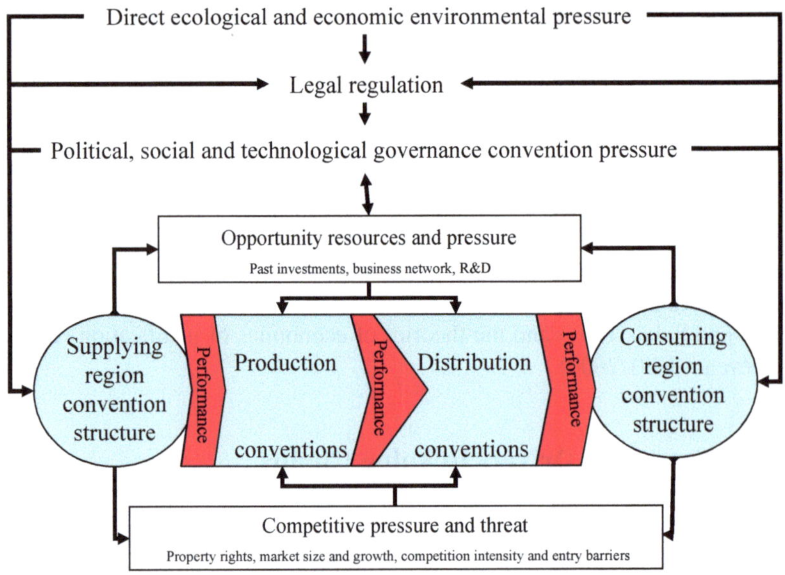

Fig. 2.1 Structure-Convention-Performance (SCP) value chain model. (after Trondsen 2012)

The theory of conventions (Salais and Storper 1992; Storper and Salais 1997; Storper 1997) takes industrial production as its main field of discussion. According to Storper and Salais (1997, p. 5) "*... industrial production is organized around the making of particular products; it is in specific product markets that competition takes place*". Nevertheless, industrial goods are not just mere objects, but "*... the material result of an extraordinarily complex ensemble of social processes* [and] *relations between persons*" (pp. 37–8). This conceptualization goes beyond the individual firm's responsiveness as an explanation for each industry's performance and turns the value chain into a substantive research topic for economic geography.

Market conventions regarding product qualities, promotion, distribution, service and prices may drive consumer demand (Bestor 2004; Wilkinson 2006; Trondsen and Young 2006). The management conventions in each of the harvesting, production and distribution stages of the value chain may be framed by performance measures like price, profit, market shares etc.

Figure 2.1 also illustrates that the embedded resources and conventions in the structural environment also influence the configuration and performance of each production stage, as well as the value chain as a whole. In fisheries for example, the business orientation and value chain configuration are to a high degree influenced by fish harvest fluctuations (Trondsen and Johnston 1998). The geographical basis of the value chain provides resources and barriers for operational, as well as competition protection and pressure. Regulation is as illustrated in Fig. 2.1, central in this discussion since conventions, informal at their origin, are often the basis for formal legal arrangements, which compel producers to follow the rules and practices considered by authorities as conducive to realization of the national values and goals of the food sector. Other formal organizations, private/public or profit/non-profit, located between producers and consumers play specialized roles related to food provisioning.

Regulation of the opportunity to enter producer or distribution strategic groups may reduce competition intensity and willingness to innovate and change. This is the case when the government issues limited licences for fishing or patents or when businesses invest in merger and acquisition to gain market dominance. Access to capital together with growth and the size of the served market relative to the total supply influence the competitive pressure in the value chain (Hayter and Patchell 2011, p. 17). The more protected the dominating participants are in each production stage, the better is their shared position in gaining monopoly profits in the transaction process without innovation (Barney 1996; Porter 1985).

The available value chain resources, convention structures and competition intensity may also lock-in and thereby limit the participant's value adding possibilities. This takes place notably in specialized value chains with high internal rivalry and high exit barriers, for example in industries embedded in outdated technologies (Porter 1985).

The competitive strength of each production and distribution stage in a chain influences also its exposure to opportunity resources and the pressure from competitors. Past investments in specialized and market-oriented technology, business networks and research and development (R&D) tend to influence strategies and access to networks, offering inspirational conventions and pressure to innovate (Grunert et al. 2010; Rogers 2003; Narver et al. 2004; Day 1994; Sousa de Vasconcelles e Sá and Hambrick 1989). The available internal and external opportunity and innovation resources and pressure or the innovative environment may thus influence each production stage, as well as the value chain as a whole (Porter 1990). Threatening competitive pressure related to the firms weekness may also occour in line with the SWOT model (strengths, weaknesses, opportunities and threats) described by Fleisher and Bensoussan (2002). The main connector between each production stage is thus the control or coordination of the industrial and market transactions that carry out the supply of products in demand, and services downstream in the value chain, against competitive purchasing payment (Gereffi et al. 2005; Grunert and Ellegard 1993).

The configuration of the value chain may therefore reflect the allocation of control over the value-adding transactions, which bind the operational production stages together. For example the Norwegian klipfish industry located in the northwest Norway has a strong influence on the salted fish value chain between the harvesting ground for codfish in the Northern Norway and the klipfish markets in Portugal and Brazil. Within this system, entrepreneurial drivers are participants in search of satisfactory rather than maximum profits, as Cyert and March (1963) and Simon (1957) have pointed out. The conventions prevailing among the units of the value chain are kinds of intangible trust assets reflecting the firms' competitiveness, which are contextually influenced (Dulsrud 2002). This may explain how the comparative advantage is developed, while the control over the value chain transactions (market power) achieved by the firm explains its actual competitive advantage (Prahalad and Bettis 1986; Porter 1985).

Adaptations to the dominant conventions take place within each stage of the chain (Gary and Wood 2011; Grunert et al. 2010). Each production and distribution stage is also influenced by different crossing value chains as well as different functional, legal and geographical contexts. Such inter-chain competition is demonstrated for instance by value chains for fresh cod, frozen cod and salted cod which are supplied from the same cod sources (Helstad et al. 2005). The participants therefore balance this inter-chain competition in a broader production system, which includes several value chains (Fisher 1997). Each stage in the chain must therefore compete to control its share of the market and input factors in the production systems. This is especially the case if they supply similar or substitute products and services as other value chains (Porter 1985). The firms involved in production and organization may thus capitalize on conventions as common platforms if they secure competitive advantages towards other competing strategic groups (Prahalad and Bettis 1986; Bettis and Prahalad 1995). The individual firms that compete within each stage of the value chain, such as primary fish processors, have common interests reflected in the behaviour, guided by the conventions shared among several similar production environments. Shared conventions within each specific stage of the value chain are thus built on common experiences and joint reflexivity (Storper 1997). This development of common, competitive convention platforms in the functional and geographical parts of the chain is one main concern of this book.

The structure of business networks between firms, for example between primary processors and retailers or between processors, may strongly influence the accepted rules of business practice among the firms and their relative control over production factors and distribution channels. The understanding of conventions by Salais and Storper's 1992, p. 174) as "...practices, routines, agreements, and their associated informal and institutional forms which bind acts together through mutual expectations" in the production realm, is in reality a conceptualization of the social competitive pressure that reflects the control of 'accepted' conduct in business networks. Such conventions or business models reflect the perceived opportunities and pressures in the firms' environment. Formal legal regulations (like access to fishing licenses), ecological conditions (like access to large salted cod) and economic environments (like access to financing of salted fish storages between the harvest season in the winter and the next year's Lenten consumption season) may influence norms or codes of conduct as illustrated in Fig. 2.1. The legal, social and ecological pressures will then influence business transactions in the value chains carried out in competitive arenas in each stage of the value chain where rivalries take place between individuals and groups.

The performance in each processing stage of the value chain is thus not only a function of the manager's relative control over transactions between the input suppliers and paying customers, but also the competitive adaptation to production and consumption conventions that are relevant for the whole chain of functions (Fisher 1997). The actors in the seafood value chains thus have to be engaged in continuous communication, controlling and reacting of opportunities, pressures, and threats in the factor and customer markets through intermediary social and regulatory networks and organizations (Trondsen 1985). This layer of organizations and networks thus indirectly binds the value chain transactions together.

Changes in society's governance rules, such as the need for general taxation, new policy, and allocation of fish quotas, as well as ecological changes in climate and fish migration, may influence the value chain's comparative advantages, and then the development of governing conventions in the value chain (Porter 1985). Changes in the ecological environment (e.g., changes in climate, fish migration pattern, or size of the fish stock) may influence value chain performance directly and indirectly (e.g. by fishing regulations aimed at preserving sustainable fish production), as illustrated in Fig. 2.1.

Governance rules include interpretation, management, and implementation of regulations in different arenas, such as courts, government administration, and other relevant social fields. Formal or informal professional or interest groups have the opportunity to influence the governance conventions relevant for their value chain. Saltfish Processors of salted fish can for example promote increases in fish quotas for coastal vessels that catch bigger cod that are more suitable for saltfish processing than for cod caught by the off-shore trawler fleet, and they can promote regulation of the use of additives in salted fish to improve product whiteness. The different governance arenas may also interact with each other as part of a power struggle to control the direction of value chain development (Trondsen 1985). The industry actors may influence their position by participation in many market arenas in addition to their primary product market. Participation in the political arena may secure fish quotas and tax reductions, in the economic arena to secure financial services, in the social arena to secure workers, in the technological arena to secure new efficient production equipment, in the ecological arena to secure fishing grounds, and finally in the legal arena to secure protective regulations. The directly related governing conventions thus unfold themselves according to their function. However, in this book, these conventions are subordinated respectively to the conventions of the production systems and consumption conventions already accounted for.

2.4 Different Kinds of Convention Approach in Food Value Chains

Three main conceptual perspectives of *institutions, values, and networks* inform the more general theoretical background of conventions that conceptualize the relationship between food value chains and the embedded society. These three concepts may explain activities in production, distribution, consumption and regulation of food. Individual producers must take decisions within the limits defined by conventions in order to supply profitable goods to consumer markets. In this view, laws and institutions are the formalized versions of the former, more informal practices. Storper and Salais have identified up to six groups of conventions related to technology, markets, participation, identity, labour, and the role of the state.

The *conventions of technology* offer solutions to availability of the material resources and knowledge processes used in the industrial realm. *Specialized* technologies are often restricted to certain locations (or 'sticky places', in the words of Markusen 1996) in the value chains. Singular raw materials or communities of technicians have framed the basis for a long-term territorial economic specialization: watches in Switzerland; software in California; red wines in Burgundy; or the Norwegian klipfish, all relying on a convention of specialized technology. *Standardized* technologies may be deployed anywhere, regardless of the tacit knowledge embedded in specialized technologies, so they are suitable for mass production in low-cost locations where economies of scale are more easily achievable (Storper 1997).

The *conventions of markets* offer solutions of product sales places and may differentiate markets demanding generics from dedicated goods. *Generic* goods are those aimed at satisfying basic common needs or demands from the bulk of consumers. Price is usually the most important criterion when purchasing these undifferentiated goods. Nevertheless, competition in the current world economy is shrinking the market share for generic goods and stretching the position of *dedicated* and differentiated goods whose brand, origin, design, designer, or any other intrinsic or extrinsic properties attached by producers enhance their attractiveness for specific niches of consumers (Porter 1990).

The *conventions of participation* give solutions for entering and operating in the industry. *Restricted* participation describes a situation where entrance barriers are high due for instance to legal surveillance over the quantity of production (quotas for fish catches or milk production in the European Union), or to the control that a community of professional experts on qualification of new trainees (i.e. by limiting access to vocational schools or university faculties). When participation is *open*, no specific barriers arise against newcomers, beyond the ordinary constraints of any business.

The *conventions of identity* refer to qualification solutions of the participants engaged in the business. Identity is a *personalized* asset that actors assess each other by their individual reputation, their particular skills, their field of expertise, or their membership in the geographical or professional communities that govern the industry. Identity is *abstract* or interchangeable when actors are qualified only by their formal diplomas or by their quantitative contribution to the production process, regardless of any more nuanced or qualitative criteria.

The *conventions of labour* deal with solutions of *productivity* and management of *employment*. Workers in any industry are supposed to earn a monthly wage by making concrete efforts to produce successful industrial goods, that is, they accept that they have to be *productive* enough for the firm to stay in business. Nevertheless, they also have to accept that poor performance or market fluctuations may lead to *unemployment* when companies adjust the working time, the size of the staff and the total wages either to its revenues or to the ups and downs of demand.

Finally, the *conventions of the state* include solutions and actors who coordinate private interests and whose goals are collective interests and will, and not individual interests, with maximization of profit or market share. Nevertheless, according to Storper and Salais (1997), the state may strive for the common will in at least three different ways. The *external* state supplies clear-cut rules and a huge amount of resources for individuals to pursue the common will, defined by the state. In the *invisible hand* fashion, the *absent* state tries to maximize an individual's opportunities to achieve their own goals and interests, with no *ex ante* definition of common guidelines apart from the maximizing of private profits. The *situated* state enhances cooperation with individuals and interest groups who have defined their joint action frameworks, and helps those groups to develop their duties and improve their economic performance.

Within the context of this book, institutions involving fishing quotas, catching and handling procedures, fish auctions, salting, freezing and other processing technologies, professional know-how and training, firm management or coastal community development are discussed as conventions that frame the aim and possibilities for coordination among Norwegian salted fish producers. Those conventions may match with conventions used by consumers to buy fish or not.

Changes in supply capabilities as well as in consumer demand preferences may, however, drive food markets built upon transactions of products and money in dynamic value chains. Currently, new concerns are influencing consumer' behaviour. Boltanski and Thévenot (1991/2006) have developed a comprehensive analytical framework to explain the *orders of worth,* which are included in negotiations and compromises in transactions among different values and interests (Wilkinson 2006). The *orders of worth* may however express the content of the conventions (Murdoch and Miele 1999; Murdoch et al. 2000; Morgan et al. 2006; Straete 2004; Trabalzi 2007).

Boltanski and Thévenot's point of departure is the different criteria we use to justify individual behaviour in different situations and contexts: the home, the workplace, the community, the public arena. They have rules of their own that we follow to differentiate between appropriate or inappropriate words, behaviours or objects. With reference to these authors, we may distinguish between seven kinds of decision situations or conventions "worlds": *industrial, market, domestic, civic, public, inspirational* and *ecological*. The *orders of worth,* as the relevant value priorities and solutions, reflect different criteria for justification.

In Salais and Storper's 1992, conceptualization, conventions are the basis for the constitution of different worlds of production and summarized under the category of *production conventions*. Boltanski and Thévenot's 1991/2006 concept of orders of worth is discussed here as covering *values* in general. In our specific analytical perspective, we see the seven kinds of justifications mostly as influences of the values and interests that underpin consumer decisions. These two main ways of organizing the use of the concept of conventions, such as two groups of production or consumption conventions in the field of food studies, may also be related to views on food production that are said to be dominant in Europe (Parrot et al. 2002). Industrial production dominates Northern Europe's food system, while the local culture obviously influences both production and consumption in southern Europe. Nevertheless, conventions are active supporting decisions both in institutions and industrial systems, as well individual preferences and, to some extent, on the moral values that predominate in Southern Europe's localized food cultures. The economic life encompasses all seven worlds of convention. Economic actors (companies, managers, workers, consumers, trade unions, regulatory bodies) do not live separate lives, but as illustrated in Fig. 2.1 have to deal with complex situations facing their internal operations, their performance in the markets, their relationships and responsibilities with the local milieu, their public reputation, their values, and their purchasing power. Accordingly, the order of worth of *industrial conventions* relies on technical efficiency and standards to deliver reliable and durable goods. The order of worth of *market conventions* focuses on trade where margins and profits (for companies) and satisfaction and usefulness (for consumers), price and wealth being its core assessment criteria. The order of worth *domestic conventions* is concerned with tradition and with trust and place as cornerstones in the justification process. The order of worth of *public conventions*, on the other hand, stresses the role of reputation (i.e. trademarks, branding, labelling) as a key driver for economic transactions. The order of worth of *civic conventions* judges economic decisions and processes according to their contribution to social cohesion, welfare and justice, while the order of worth of *ecological* conventions is mostly concerned with their implications for environmental sustainability. Finally, the order of worth of *inspiration* conventions is important for innovation because of the commitment of individuals with beliefs, ideals or missions.

The coordination of economic life is mainly based on industrial and market orders of worth conventions, but "... the complex societies that we are studying cannot be confined within any one of the worlds that we have identified" (Boltanski and Thévenot 1991/2006, p. 195). Therefore, producers must be aware that markets are imperfect, that consumers are not purely calculative devices, and that a number of different qualities may be attached to industrial goods because consumers are also (and increasingly, at least in developed countries) sensitive to domestic, public, civic, ecological, and even inspirational values. Food is particularly appropriate as a field for developing strategies other than volume and price, because of its material and representational connections with nature, health, justice, culture, tradition, taste, or place.

Different orders of worth conventions are thus materialized in consumer solutions and decisions, and influence producers' management and practices and embodied in labels, logos, packages, advertisements, discourses and any other marketing devices, in order to enhance the qualities, attributes, features, and traits corresponding to different common principles for justification. This book discusses the ability of Norwegian salted cod producers, following northern European production conventions, to meet new demands stemming from the particular mix of orders of worth conventions that prevail in the Spanish seafood market.

The third conceptual perspective in our discussion relates to formal and informal networks and their intermediary role in the food value chain as responsible for matching production and consumption conventions as illustrated by the arrows in Fig. 2.1. At the informal level, parties collaborate on various forms of strategic coalitions where attempts can be made to realize joint intentions and goals. Nevertheless, organizations are formal coalitions of human, technical and knowledge resources that have a defined goal or purpose. Companies, trade unions, and producer associations usually pursue private interests. National, regional and local authorities, along with public agencies with specific goals, are supposed to represent a common will, according to the conventions of the state previously defined. For both types of organizations, laws, norms, rules, and acts are the normal tools for codifying practices and values in a dynamic bargaining process that, at any time and location, represents the power and influence of each actor in the value chain (Lindkvist 2010).

Table 2.1 Conceptual framework for the socio-economic analysis of the institutions, values and networks in Norwegian-Spanish salted fish value chain

Value chain STRUCTURE	Demand, factor resources and organization
Velue chain NETWORK	Markets (for whom?) suppliers (from whom) technology (how?) participation (who?) identity (who?—where?)
Formal value chain CONVENTIONS	Laws and guidelines
Informal value chain CONVENTIONS	Goals, norms, rules, attributes, features, traits, expressions behaviour, decisions, practices, orders of worth
Competitive value chain ENVIRONMENT	Product markets, raw material market, industrial and commercial factor markets state, labour, companies & traders, trade unions, associations R & D agencies, domestic, public, civic, ecological, inspiration (innovation) conventions innovation systems and authorities (any scale), law
Link between the environment and the (food) value chains	Auctions, quotas, shares, catching & handling, family, community. marketing organisation, labels, logos, packages, discourses, advertisements, distribution and prices

Table 2.1 summarizes our discussion and is meant to underline the fact that all goals, norms and value priorities, displayed in the different situations and within the different networks, are operationalized as conventions. The following table summarizes how the three perspectives that have been referred to in the convention debate (and linked to institutions, values and networks) may be systematized to create a consistent convention concept.

2.5 Convention Variation Along the Value-Chain

The discussion so far has demonstrated that the conventions linked to the value chain do not operate in a void. Each link in the value chain represents a market. These markets are institutions facilitating the exchange of price-based goods and services. Marketing conventions are characterized by competition patterns between the actors in the market and the institutional networks in the actors' local environment. However, markets are also spatially abstract and place-specific at the same time.

The discussion has also shown that the value chain of connected production stages defined by rules, social practices and governance structures are influenced by the specific situations and locations that characterize the chain itself (Ponte 2009; Sánchez et al. 2010). In addition to this, food products from specific regions often identified by their regional characteristics are important traits of the food market in Mediterranean countries (Parrott et al. 2002; Díaz and Gómez 2010).

The core question will then be how to investigate chain units, conventions and geographical locations in their totality, and which of these three main elements represents generative causal factors for the success of the trade activities running through the (salted fish) value chain.

Driving forces in the economic development of firms often have a geographical location as a starting point. Path-dependent socio-economic processes in places or regions as networks of places may result from the specific history of those processes (Coe 2011, p. 84). Still "... we need to understand regional 'lock-in' as a multi-scaled process, and one which has a high degree of place-dependency..." (Coe 2011, p. 84). Many natural resources are in specific locations, and not easily relocated.

Many food resources such as certain fish species have their localization in geographically bounded fishing grounds, with boundaries determined by their migratory patterns. The exploitation of fish resources has influenced specific local cultures and their adjacent location of economic activities.

There is a relationship between fishing places and their dependent fish resources, and the consumer regions' dependence on their own cultural background, linked to the consumption of fish products. The main path dependent driver for salted codfish is Catholic tradition, which demands a protein substitute for meat in the Lent season (see Richter-Hanssen, Chapter 3). People in these countries are to some extent locked in the local culinary culture and consumption patterns. This does not mean that the links remain unchanged. Local manufacturers and retailers therefore have established a long tradition of trading salted seafood in their markets, which is somewhat different from one part of Spain to another. Socioeconomic practices and values related to food have thus resulted in many heterogeneous, geographic markets for salted fish in Spain.

This book discusses these causal relationships between the availability of fish resources, fishing territories or communities and conventions in value chain and consumer markets. Specific conventions will be relevant for each production and place-based function. Clearly, domestic production conventions are relevant in the fishing areas, as well as in the processing areas, whether they are located in Norway or in Spain. Different specific geographical conventions will persist and influence different content and each type of convention, for example when the focus is shifted from resource appropriation to value-added production and consumption (Lindkvist and Sánchez 2008).

The contemporary economic geographical approach acknowledges that the specific context, consisting of an accumulation of activities and institutions in a specific place, is a generative causal mechanism. These specific links then determine which conventions are in focus and how they enhance or hamper competiveness and innovative ability in the specific area in question.

Figure 2.2 depicts some of these conventions as general parts of this regulatory framework, but with an unequal effect on the value chain units. In turn, these different outcomes arise from the composition of the regulatory arrangements of the different geographical and functional parts of the value chain.

Figure 2.2 highlights the value chain space and conventions as analytical categories for study of the Norwegian–Spanish salted fish trade. The two large spatial levels consist of Norway (as the supplier market) and Spain (as the consumer market) for Norwegian fish. As each of these two national territories comprises several production regions and consumer regions, a further territorial breakdown would result in an even more detailed map, with smaller regions. This would mean dividing the different links of the value chain according to localized stations for the catching and

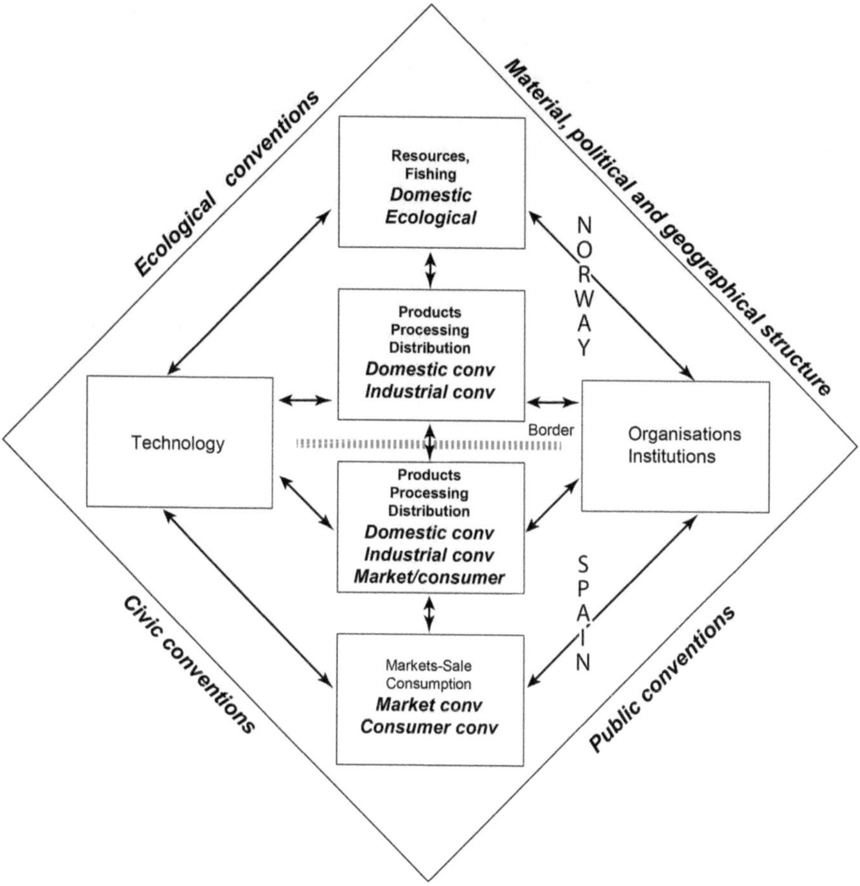

Fig. 2.2 The salted fish value chain and the institutional context

production of salted fish in different parts of Norway, and the production, adaptation and consumption of salted fish in various Spanish regions.

However, we could question whether division into even smaller spaces of the chain has any analytical value, at least in Norway. For instance, if the regulatory framework prevents local culture from innovating, the more uniform becomes the entire value-chain. However, local technological and organizational conventions may still affect the performance of actions. The technological conventions influence production conventions, especially the industrial, while the organizational conventions influence development of networks and institutional development. These convention categories frame the operation of the production systems in the value chain. In addition, such systemic conventions will most likely be part of the accumulation of experiences and social capital, which, together with the accumulation of physical capital, characterize the different regions.

Consumer conventions, with their expression of values and preferences, may give solutions and justify the choices taken by other actors in different stations of the value chain. The domestic conventions will for instance, give solutions and act as justifications of the standpoints taken by the actors in the different environments along the chain; how they will produce, and how they will protect their own interests in the value chain (Lindkvist 2010). The same justification process applies to activities in domestic systems and in the consumer parts of the value chain.

Some of the order-of-worth based conventions such as civic conventions, ecological conventions and public conventions, function as frameworks for the total production chain. The conventions used to categorize typical situations are not static, and we cannot easily predict how we will understand them in the future. Many situations or conditions problematize adaptation within different local production systems. For example, consumer demand for products, as tailored to production conventions, may change. Some producers may wish to, or even be forced to, follow new trends, while others may refrain from doing so, or for other reasons be unable to change. Different actors within the same local production systems may even interpret the information they receive differently (Rogers 2003). This may for different reasons put pressure on the agreement on conventions between the production units, both horizontally and vertically. For such reasons, production units and the governing rules or conventions participating in the dynamic processes of change may transform over time.

This book investigates why the Norwegian salted fish industry has not established itself as part of the value chain in Spain. For the most part, if the producers in Norway are to be generally successful in another country, the conventions of the value chains in both countries must match. This is important because if the conventions of the Norwegian commodity producer cannot change to fit consumer markets, the Norwegians will need to establish alternative value chains that will allow their conventions to unfold in the importing country. This will be possible through the existence of alternative value chains based on different products and market conventions, which already find resonance among producers.

2.6 Summary and Conclusion

This chapter has outlined the analytical concepts and models in Economic geography to be used in this book. Identification of structure, the drivers and the governing conventions over time in the value chain, are the main factors to be understood. The value chain and its drivers are understood in the context of the Structure-Convention-Performance (SCP) model. We argue that the governing conventions in the Norwegian salted cod export to Spain, especially as a platform for any future attempt to recapture this market, has to be understood in a broader and deeper sociological context, related to different kinds of institutions, networks, values, and qualities. The differences in the interface between the value chain and the spatial and contextual properties and the conventions commonly used in food trade literature, should receive special attention. Focus is on discussion of empirical issues regarding history, prices, exports, companies, consumption, lock-in, education, or coordination.

References

Barnes TJ (2000) Inventing economic geography, 1887–1960. In: Sheppard ES, Barnes TJ (eds) A companion to economic geography. Blackwell, Oxford

Barnes TJ (2001) Lives lived, and lives told: biographies of geography's quantitative revolution. Soc and Space: Environ Plan D 19:409–29

Barnes TJ, Farish M (2006) Between regions: science, militarism, and American geography from world war to cold war. Ann Assoc Am Geogr 96:807–26

Barney JA (1996) Gaining and sustaining competitive advantage. Addison-Wesley, New York

Bestor TC (2004). Tsukiji. The fish market at the center of the world. University of California press, Los Angeles

Bettis R, Prahalad CK (1995) The dominant logic: retrospective and extension. Strateg Manag J 16:5–14

Boltanski L, Thévenot L (1991/2006) On justification: economies of worth. Princeton University Press, Princeton

Coe NM (2011) Unpacking globalization: changing geographies of the global economy. In: Leyshon A, Lee R, McDowell L, Sunley P (eds) The Sage handbook of economic geography. Sage Publications, London, pp 89–101

Cyert RM, March JG (1963) A behavioural theory of the firm. Prentice Hall Inc, New Jersey

Day GS (1994) The capabilities of market driven organizations. J Mark 58(4):37–52

Díaz-Méndez C, Gómez-Benito C (2010) Nutrition and the Mediterranean diet. A historical and sociological analysis of the concept of a "healthy diet" in Spanish society. Food Policy 35(5):437–447

Dulsrud A (2002) Tillit og Transaksjoner. En kvalitativ analyse av kontraktsrelasjoner i norsk hvitfiskeksport. National Institute for Consumer Research, Report 2, Oslo

Fisher ML (1997) What is the right supply chain for your product? Harv Bus Rev 75(2):1005–1116

Fleisher C, Bensoussan B (2002) Strategic and competitive analysis. Methods and techniques for analyzing business competition. Prentice Hall, New York

Foster J, Muellerleile C, Olds K, Peck J (2007) Circulating economic geographies: citations patterns and citation behaviour in economic geography, 1982–2006. Trans Inst Br Geogr 32(3):295–312

Gary MS, Wood RE (2011) Mental models, decision rules and performance heterogeneity. Strateg Manag J 32:569–594

Gereffi G, Humphrey J, Sturgeon T (2005) The governance of global value chains. Rev Int Political Econ 12(1):78–104

Grunert KG, Ellegaard C (1993) The concept of key success factors: theory and method. In: Baker MJ (ed) Perspectives on marketing management. Wiley, Chichester, pp 245–274

Grunert KG, Trondsen T, Campos EG, Young JA (2010) Market orientation in the mental models of decision makers. two cross-border value chains. Int Mark Rev 27(1):7–27

Hayter R, Patchell P (2011) Economic geography: an institutional approach. Oxford University Press, London

Helstad K, Vassdal T, Trondsen T (2005) Price links between auction and direct sales of fresh and frozen fish in North Norway (1997–2003). Mar Resour Econ 20:305–322

Lindkvist KB (2010) Mistrust and lack of market innovation: a case study of loss of competitiveness in a seafood industry. Eur Urb Reg Stud 17(1):31–43

Lindkvist KB, Sánchez JL (2008) Conventions and innovation: a comparison of two localized natural resource-based industries. Reg Stud 42(3):343–354

Markusen A (1996) Sticky places in slippery space: a typology of industrial places. Econ Geogr 72(3):293–313
Murdoch JT, Miele M (1999) Back to nature': changing 'worlds of production' in the food sector. Sociol Ruralis 39(4):465–483
Murdoch J, Marsden T, Banks J (2000) Quality, nature, and embeddedness: some theoretical considerations in the context of the food sector. Econ Geogr 78:107–125
Morgan K, Marsden T, Murdoch J (2006) Worlds of food. Place, power, and provenance in the food chain. Oxford University Press, Oxford
Narver JC, Slater SF, MacLachlan DL (2004) Responsive and proactive market orientation and new-product success. J Prod Innov Manag 21:334–347
Parrott N, Wilson N, Murdoch J (2002) Spatializing quality: regional protection and the alternative geography of food. Eur Urban Reg Stud 9:241–261
Ponte S (2009) Governing through quality: conventions and supply relations in the value chain for South African wine. Eur Society Rural Sociol 49(3):236–257
Porter ME (1980) Competitive strategy. The Free Press, New York
Porter ME (1985) Competitive advantage. The Free Press, New York
Porter ME (1990) The competitive advantage of nations. Macmillan Press, New York
Prahalad CK, Bettis RA (1986) The dominant logic: a new linkage between diversity and performance. Strateg Manag J 7:485–501
Rogers E (2003) Diffusion of innovations. The Free Press, New York
Salais R, Storper M (1992) The four worlds of contemporary industry. Cambridge J Econ 16: 169–193
Sánchez JL (2003) Naturaleza, localización y sociedad. Tres enfoques para la Geografía Económica. Ediciones Universidad de Salamanca, Salamanca, pp 257
Sánchez JL, Aparicio-Amador J, Alonso-Santos JL (2010) The shift between worlds of production as an innovative process in the wine industry in Castile and Leon (Spain). Geoforum, 41:469–478
Scott P (2000) Social network analysis: a handbook. Sage Publications, London
Simon H (1957) Administrative behavior. The Free Press, New York
Sousa de Vasconcellos e Sá JA, Hambrick DC (1989) Key success factors: test of a general theory in the mature industrial-product sector. Strateg Manag J 10:367–382
Storper M (1997) The regional world. territorial development in a Global Economy. The Guiford Press
Storper M, Salais R (1997) Worlds of production. The action frameworks of the economy. Harvard University Press, Cambridge
Straete EP (2004) Innovation and changing 'worlds of production': case studies of Norwegian dairies. Eur Urban Reg Stud 11(3):227–41
Trabalzi F (2007) Crossing conventions in localized food networks: insights from southern Italy. Environ Plan A 39(2):283–300
Trondsen T (1985) Industriell Innovasjon. En studie av institusjonelle forutsetninger. Dr.scient. thesis, Institute for fishery science, University of Tromsø
Trondsen T (2012) Value chains, business conventions, and market adaptation: a comparative analysis of Norwegian and Icelandic fish exports. Canadian Geogr/Geographe Canadien 56(4):459–473
Trondsen T, Johnston RS (1998) Market orientation and raw material control. J Mark Focus Manag 3:193–210
Trondsen T, Young JA (2006) The role of fish auctions in value adding in fish marketing chains. In: Ashe F (ed) Primary industries facing global markets. The supply chain for Norwegian food. Universitetsforlaget, Oslo, pp 393–427
Trondsen T, Helstad K, Young JA (2003) Market-oriented regional fisheries management—an analysis of four fish regions in the North Atlantic. Ocean Coast Manag 46:917–941
Wilkinson J (2006) Fish: a global value chain driven onto the rocks. Sociologia Ruralis 46(2) 139–153 April

Chapter 3
The History of the Norwegian—Spanish Salted Fish Trade

Einar Richter-Hanssen

Abstract This chapter investigates the historical origin of the structure of the Norwegian-Spanish salted fish value chain, and focuses especially on historical bases of the current conventions which dominate the value chain. Between 1665 when the first traces of this trade were noticed, five specific trade phases have been documented. The salted fish trade between Spain and Norway started with the Mecantile phase, and developed through liberal and freer trade phases, then the state regulated the trade again from the 1st world war until a freer trade regime took over in the 1960s. Throughout the last two centuries the challenge to keep a satisfying product quality has been met in different ways.

Keywords Historical origin · Trade phases · State regulations · Quality demands · Clip fish

3.1 Introduction

All conventions have a historical fundament and past dependent. This chapter investigates the origin of the value chain structure and the current reigning conventions in the Norway and Spain salted fish trade. Salted fish has had an important place in Spanish culture since pre-Roman times (Lindkvist et al. 2008). Tuna fish were salted and dried during the hot Mediterranean summers. After a decline in salted fish production during the period of Muslim domination, the tradition re-emerged in the thirteenth century, and from around 1500 focused on cod. Salted and dried cod grew in importance during special peak times in the year, such as Lent or Christmas, and was firmly interlinked to Catholic precepts. The industrial foundation of Norwegian salted fish exports was Spain's long-standing traditional use of salted fish, primarily cod, and the fact that Spain had long been the most important market in Europe (Wallem 1893; Vollan 1956). Until 1970 the export consisted of clip

E. Richter-Hanssen (✉)
University of Bergen, Bergen, Norway
e-mail: richterhanssen@gmail.com

K. B. Lindkvist, T. Trondsen (eds.), *Nordic-Iberian Cod Value Chains*,
MARE Publication Series 8, DOI 10.1007/978-3-319-16405-2_3

fish, i.e. the dried and salted variety. [1]From then on the consumption of clip fish in Spain stagnated, with more and more wet salted fish while the clip fish market was marginal. On its side, Norway had enormous fish resources and was in a position to meet Spain's increasing demand (Table 3.1). As a mass-producer and through Spanish trade and toll policy, Norway became a dominating player in northern Spain and Catalonia. It was a deliberate strategy by the Norwegian exporters, especially in the period from the 1830s to the 1930s, to target lower income consumer groups like the urban workers and the rural working population where price[2] was a decisive competitive factor (Wallem 1893; Vollan 1956; Solhaug 1983; Puerta Rueda 1998; see also Tables 3.2 and 3.3 in Attachment). The export focus on large quantities and low prices in preference to quality (Rathke 1907; Strøm 1949; Jensen 1963) contributed to the view that Norwegian clip fish was considered all over as a product of lower quality than clip fish from Iceland which was preferred by the more affluent urban population first in Catalonia, then in Northern Spain (Vollan 1956). As long as Norwegians could compete regarding price, one was guaranteed market allocation. When demand decreased in Spain around 1900 at the same time as quality standards gained importance, Norway lost market share.

In the following sections, I will discuss the general assertions I have outlined above in more detail. Figure 3.1 refers to the most important ports of trade in salted fish between Norway and Spain.

3.2 1665–1814: The Mercantile Phase

The first documented case of clip fish export dates back to 1665 in Trondheim (Berg 1927). The fish was most likely sent to Spain as Dutch tradesmen, who dominated the foreign trade in Europe, were also very active in Trondheim. They exported self-produced clip fish to Spain via Amsterdam, which at that time was the only market for clip fish of any significance. The first documented case of direct trade between Norway and Spain was in 1692, to Bilbao (Berg 1927; Johnsen 1942). In accordance with mercantile trade policy the dual monarchy of Denmark-Norway encouraged first Dutch, then Scottish merchants and skilled workers from among others Spain, to settle in the Møre region. They in turn taught these skills to the Norwegians.

---

[1] Clip fish is dried cod produced primarily from cod. But also other white fishes like ling, cusk and pollock. The fish is gutted, washed and stained with 2/3 of the spinal column removed. Then the fish is washed carefully again, so that all impurities are removed. Then the fish is salted and placed in stacks. The fish will then mature after 3–4 weeks. After the fish is matured, the salt is washed out of it. Previously, the clip fish was dried on rocks. Now all dried cods are produced in blowing heat ducts, where the fish is laid on the floor wagons with string bottom. The wagons run in and out of the drying canal. When the clip fish has finished drying, it is stored in cold storages (Pedersen 1992).

[2] The sources used in this chapter have with one exception (Table 3.2 in Attachment) not given the Norwegian export prices to Spain in numbers, they are only referred to as either higher or lower than the competitors.

Table 3.1 Export and value of salted fish from Norway to all markets. Export and value of salted fish to Spain. Value of export to and import from Spain of all commodities. (Sources: Wallem 1893; Blich 1928; Gerhardsen and Gertenbach 1949; Vollan 1956; Statistisk Sentralbyrå 1969; NOU 1972; Solhaug 1983; Lindkvist and Hauge 2000; Lindkvist et al 2008)

Year	Export to all markets in metric tons[a]	Export to Spain in metric tons[a]	Export to Spain expressed in percentage of total export	Value of export to all markets in 1000 NOK[a]	Value of export to Spain in 1000 NOK[a]	Value of export to Spain of all commodities in 1000 NOK[b]	Value of import from Spain of all commodities in 1000 NOK[b]
1665	0.6	0.6	100				
1750	265	265	100				
1760	1094	1094	100				
1820	2392						
1830	7448						
1840	9585						
1850	11,810	8467	71.69				
1860	18,617	14541	79.10				
1870	24,735			8614	6208	6658	847
1880	52,962			12,304	8883	9455	929
1890	55,572	35,723	61.29	17,785		11,974	645
1894	64,550	42,040	65.12				
1900	32,358	21,142	63.33	13,590		11,248	3247
1910	36,341	13,272	36.52	17,744		9081	3345
1920	36,751	15,595	42.43	43,803		34,221	38,441
1930	45,945	8255	17.96			12,214	19,344
1940	19,249	43	0.22			1714	5861

Table 3.1 (continued)

Year	Export to all markets in metric tons[a]	Export to Spain in metric tons[a]	Export to Spain expressed in percentage of total export	Value of export to all markets in 1000 NOK[a]	Value of export to Spain in 1000 NOK[a]	Value of export to Spain of all commodities in 1000 NOK[b]	Value of import from Spain of all commodities in 1000 NOK[b]
1950	37,608	4595	12.21			54,307	53,519
1960	30,021					49,328	112,031
1970	95,000	900					
1991		3163					
1996		12,983					
1999	117,049	8211					
2005		5650					
2007		7923			356,042		
2008		6342			293,293		
2009		5318			181,833		

[a]Until 1970 the export numbers included only clip fish. From 1970 on the numbers also include wet salted fish and salted filet

[b]Through the nineteenth century and the first decades of the twentieth century the export from Norway to Spain consisted in practice of only one commodity, clip fish. Other commodities, foremost timber, played an insignificant role

Table 3.2 Yearly average prices of clip fish on the regional Spanish markets of Bilbao and Barcelona 1890–1892. (Source: Wallem 1893)

Year	Bilbao in NOK pr. 50 kg	Barcelona in pesetas pr. 40 kg
1890	25.78	24.5
1891	29.99	36.6
1892	27.82	38.6

Table 3.3 Yearly average Norwegian export prices of clip fish 1820–1925 to all markets. Prices in Norwegian *kroner* and clip fish in *våg* (= 18.52 kg) 1820–1880 and in *vekt* (= 20 kg) 1890–1925. Prices 1820–1880 are converted from Norwegian *skilling* to *kroner*. (Sources: Wallem 1893; Blich 1928; Solhaug 1983)

1820	3.90
1830	3.10
1840	6.10
1850	2.80
1860	4.80
1870	5.90
1880	4.20
1890	6.40
1900	8.40
1910	9.60
1915	16.40
1920	23.80
1925	26.00

During the first decades, trade with Spain was insignificant. But from the 1730s there was a steady growth, and the importance of the Spanish market lead to an industrial transformation from stock fish to clip fish production. This process was, according to Johnsen (1942), one of the most important phenomena of the last centuries, not only for the Norwegian fishing industry, but within Norwegian economic history as a whole.

Population-wise, Spain was a large nation with approximately 9 million inhabitants in 1768. With a steadily growing population and no production of its own, the sales potential here was very good. The Norwegian markets were in northern Spain and Catalonia while the southern Spanish market preferred clip fish from Newfoundland and New England.

In the 1790s for instance with an export of 4000 metric t the Norwegian delivered at a lower price than the French in the northern Spanish market. Norwegian clip fish was also cheaper than fish from Newfoundland in Catalonia, and Barcelona "was almost wholly supplied from Norway" (Innis 1954, p. 299). When Spain raised the tariff on clip fish in 1785, this too must have affected the more expensive fish from Newfoundland most harshly.

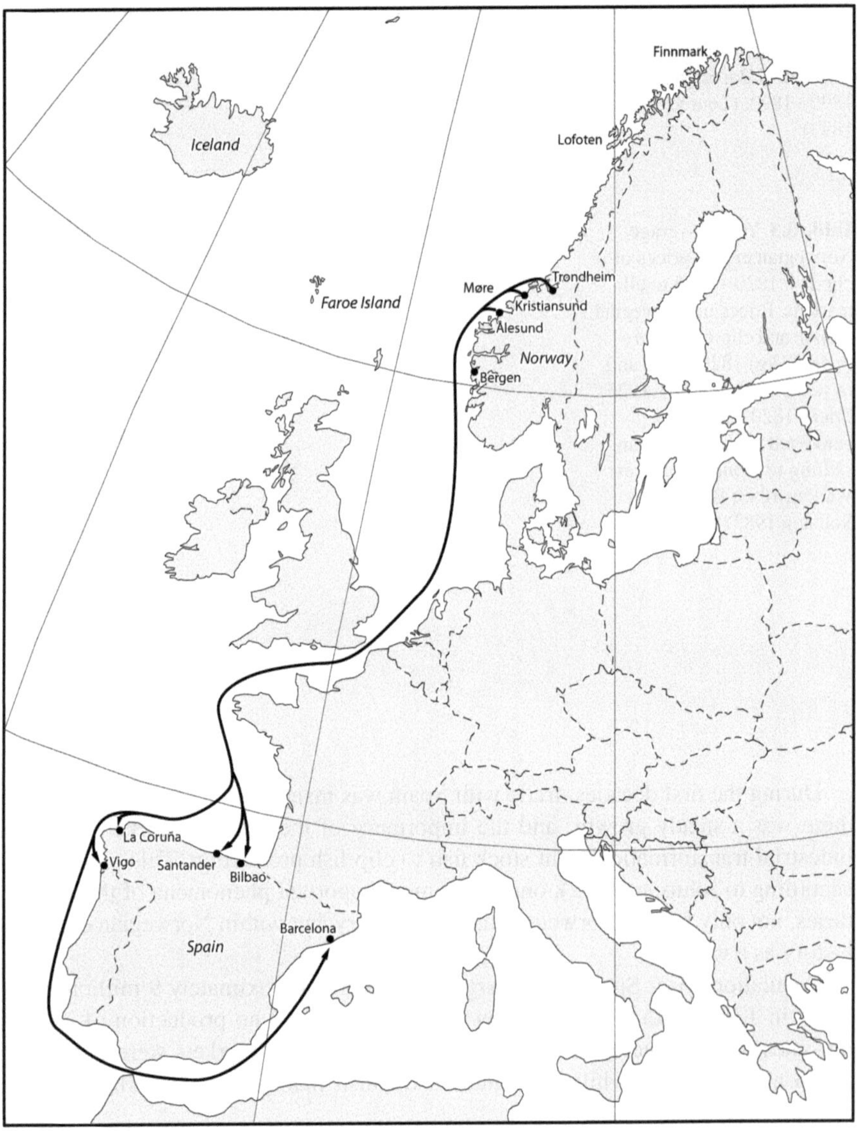

Fig. 3.1 Important ports in Norway and Spain and the trade routes between the two countries

In the eighteenth century, however, Norway was a minor actor in the Spanish market, the larger ones being New England, Newfoundland and France.

Still, around 1800 the prospects for Norwegian clip fish export appeared to be good, but when Denmark–Norway became involved in the Napoleonic Wars in the period 1807–1814, the trade suffered a serious setback that took years to overcome.

3.3 1814–1871: The Late Spanish Mercantile and Norwegian Economic Liberal Phase

The Norwegian expansion commenced in the 1820s. Norway became one of the world's largest producers because the country had large and relatively easily accessible fish resources (Berg et al. 1988). At the end of the 18th century, when the fisheries in the Møre region could no longer supply a sufficient quantity of raw goods, producers turned instead first to Lofoten and then Finnmark from the 1850s.

The significant growth of Norwegian exports was primarily due to the successful sale of large quantities of clip fish to Spain and only to that country (Wallem 1893; Solhaug 1983). "The most remarkable feature of the nineteenth-century market was the steady rise of Norwegian cod, which ultimately displaced Newfoundland cod as the preferred article" in Bilbao (Piquero and López 2006, p. 198).

In accordance with its mercantile trade policy, Spain introduced a state-controlled wholesale purchasing monopoly in 1824. However, that resulted in a near total collapse of the import of clip fish. The monopoly was therefore dissolved in 1825 and the duty on clip fish was also reduced. On the other hand, Spain introduced a so-called 'differential duty' in 1825 to guard its own shipping industry. This was a protectionist measure, with a lower duty being placed upon fish carried in Spanish ships than that transported in foreign vessels (Vollan 1956; Solhaug 1983). From 1784, Norway had departed more from mercantilism than most other countries in Europe, and from 1814 pursued a clear economic liberalistic policy (Vollan 1956; Dyrvik et al. 1979). Then Norway could not in principle respond with restrictions on Spanish import goods. In reality, except for salt, Norway imported very little from Spain in any case. The trade between the two countries benefited Norway unilaterally. (See also Table 3.1 in Attachment).

Extensive smuggling was able to partly evade the differential duty. Smuggling represented e.g. in 1866, about one-third of the legitimate trade in Barcelona (Vollan 1956).

The differential duty made it unprofitable for Norwegians to carry clip fish to Spain, and so Spaniards took over the carrying trade (Bugge 1923; Moltu 1932; Johnsen 1942; Vollan 1956). The Møre region was the leading Norwegian clip fish producer, but both Trondheim and Bergen held leading positions in exports until the 1830s. However, from the mid-1800s, as a direct result of Spanish trade policy, Kristiansund and Ålesund in the Møre region, once more took a leading export position. The Møre region had an excellent position in relation to the sea close to fishing grounds. Spanish merchants favoured these production centres as they could pick up the fish sooner here than in either Bergen or Trondheim, and then return earlier to Spain. Since both Trondheim and Bergen had several large shipping companies, they did not profit from the differential duty. Kristiansund and Ålesund did not have their own shipping fleets and thus was not affected by the duty (Op. cit.).

Although the differential import duty applied to all clip fish from all exporting countries, it gave Norway an advantage over its competitors. When the Spaniards themselves carried the fish, the trip to Norway was far shorter than the journey to, for instance, Newfoundland. Vollan (1956) argued that the differential duty was the Norwegians' best weapon against their competitors, and more so when the Norwegian fish was not always of competitive quality.

By selling the fish directly against cash payments to the Spaniards that came to Norway, the Norwegian merchants had no transport risk even if profits were lower. They could also deliver as much clip fish as they wanted. Having Spaniards buying eased the capital acquisition problem for many smaller Norwegian merchants who wanted to participate in trade, as they did not have to commit capital to exports, and could concentrate on buying the fish fresh and on manufacturing.

The best-paid clip fish was that which first arrived on the market in Spain. The timing question played such a large role that the Spanish merchants could come to Møre several weeks before the fishing started and only half-load the ship to be quickly turned around (Vollan 1956). But this could also lead to lower quality, since drying time thus was forced. In 1871 Norway and Sweden (united 1814–1905) signed their first trade treaty with Spain; here the salted fish trade played a significant role. From then on Norway could sell and transport the fish at the same rates as the Spanish. For Spaniards, it was henceforth economically unprofitable to have ships lying idle for weeks in Norwegian ports to wait for the clip fish, and they stopped sailing to Norway (op.cit.). In 1872, the differential duty was lifted.

3.4 1871–1914: Free Trade Phase

Bilbao and Barcelona were gateways for the big Norwegian clip fish markets in the Northern and Eastern Spain, respectively (Piquero and López 2006). Overall, the clip fish was a popular dish in all social strata in Spain, and had been for a long time. Clip fish was, for instance, common food in many households in Catalonia in the last part of the eighteenth century, and "even during meat days" (Op.cit.:203). Consequently, the consumption of clip fish occurred on other days than during Lent.

The largest consumers of Norwegian clip fish were the rural population and the urban working class, and on whom the Norwegian exporters concentrated. Norwegian fish sold, for example, best in the Basque Country "due to its reasonable price, and was therefore most popular with the working class" (de la Puerta Rueda 1998, p. 199).

Although price is an important factor, it seems doubtful that anyone would prefer an inferior product when it comes to taste. It must primarily have been the customers' ability to pay. As a Norwegian clip fish importer in Bilbao, put it: Price was the sole reason for the sales of inferior qualities (Clausen 1927). This is confirmed by the conditions in Catalonia which was the most demanding of the clip fish markets in Spain, and the one that paid the best. It was one of the most prosperous regions due to an early industrial development (Fougner 1941; Vollan 1956).

In the larger cities and among middle and upper classes Norwegian fish was only used to a lesser extent. They preferred the Scottish and Faroese fish that was of better taste (Wedel Jarlsberg 1932). The huge purchasing power in Bilbao: "not only enhanced the cod trade, but also determined its destination, with the citizens of the port choosing to consume the higher quality Scottish and Danish [Faroese] cod, while lower quality produce was transported inland" (de la Puerta Rueda 1998, p. 200). Icelandic clip fish was also highly prized in Bilbao.

The Norwegians were so confident of their market position that although they got signals that Norwegian clip fish lost out to the Icelandic fish, in around 1900, they did not realise the danger (Vollan 1956). They believed that Icelandic fish was sold to the high-quality demanding middle and upper classes, while the Norwegian fish went to people of humble means, and that the quality was of secondary importance (Op.cit.).

The same conservative attitude towards market adjustments is evidenced again and again, also in markets other than the Spanish market (Vollan 1956). This attitude towards the Spanish market, that developed even as Norwegian exports were growing, would prove to have dire consequences in the long term.

At the same time, it is easy to understand how the Norwegians could be nonchalant, because the Spanish market grew throughout the 1800s and peaked with the considerable quantity of 42,040 metric t in 1894, which is similar to 153,000 t of live fish. The Norwegian export of clip fish to Spain has never, in later years, come even close to these numbers.

From the latter decades of the 1800s, the Spanish market could place higher demands on quality standards as more competitors joined the game. Iceland did not produce the greatest amounts, but they became a competitive threat because they concentrated on one market—the Spanish. The same applied, to a certain degree, to Scottish clip fish. The Faroe Islands, France and all the North American producers ramped up activity. Directly after 1900, England, Germany and Spain itself joined the action (Fiskets Gang 1911; Fiskets Gang 1914; Vollan 1956).

After 1873, the fisheries in Spain were liberalised and modernised, with a rapid growth in the quantity caught and in the consumption of fresh fish. Until 1910, a lack of train capacity limited domestic fish exports from the many small fishing ports, but as railroads were expanded so too did the availability of fresh fish in inland regions (López 1999).

Clip fish production occurred in Spain as late as 1908, based on hauls of corbina along fishing banks near the Canary Islands (Fougner 1941). But a greater production of clip fish, based on fish from the Newfoundland banks, did not occur in Spain before 1927 when the first large ocean-fish company, based in Pasajes, was founded (Norges Handels- and Sjøfartstidende 1928).

After 1894, the Norwegian export trade showed a declining tendency, first in real numbers and then relatively (Blich 1928; Fougner 1941, Table 3.1). All of the negative factors that have been discussed, and which waited like a ticking time bomb under the trade conditions, finally exploded. Norwegians lost hold on its markets one by one.

The first market Norway lost was Catalonia, which was known to be the world's most particular. The decline started in the 1890s, a result not of a change in taste or need among consumers, but, first and foremost, because of sloppy processing and ambivalence toward processing on the part of manufacturers (Fiskets Gang 1911).

By 1910 the Norwegian exports to Barcelona was but a shadow of its former self and constituted e.g. only five percent of Iceland's share (Fiskets Gang 1914).

Buyers in Catalonia had complained about Norwegian fish for years because of poor processing, meaning fish kept poorly and were afflicted with clip fish mould on arrival (Fiskets Gang 1911). The consulate in Barcelona had briefed exporters on such issues, but complaints were not followed up. In Catalonia, quality played a large role; fish that was not absolutely white and firm could not easily turn a profit (Op.cit.).

Consumers in Barcelona desired a large, plump, white and mould-free fish with a lower salt content and less dryness than the Norwegian fish (Vollan 1956). As opposed to the Norwegians, Icelanders geared their production toward meeting these demands. They also used salt from Ibiza, which gave the fish a white colour and protected it better from moulding. Norwegians, on the other hand, applied the Italian Trapani salt which they traditionally used for herring, and transferred their salting practice from herring to cod. But, since the Trapani salt contained sulphur, it turned the fish yellow (Op.cit.).

3.5 A Choice between Quantity and Quality

To explain this downfall we have to look first and foremost at the above-mentioned problem with poor quality. Through the history of the Norwegian–Spanish trade, complaints of the poor quality of Norwegian fish have been a persistent problem (Deinboll 1839; Rathke 1907; Eidsvaag 1921; Bugge 1923; Moltu 1932, Vollan 1956; Tande 1957). Other more time-limited factors like increasing competition from other export countries, increasing Spanish fishing, and attitudes toward customers and a lack of good business practice have played a somewhat lesser role. I will return to these factors later on.

It would be incorrect to allege that Norwegian fish was poor throughout and at all times. In great contrast, it was often preferred ahead of other fish, and competitors also faced similar criticism from Spanish buyers. Moreover, there were differing qualities of Norwegian goods. But, on a general basis, the clip fish was considered to be inferior with regard to Scottish, Icelandic and Faroese products. The Norwegian clip fish was on the other hand, regularly estimated to be better than French fish and could remain competitive with that from Newfoundland (Vollan 1956; Puerta Rueda 1998). Such judgments also varied according to taste. Both French and North American fish were more watery than their Norwegian counterpart.

The first documented complaints appeared in the 1790s when the natural scientist J. Rathke (1907) travelled around in Norway at the state's expense to evaluate the fisheries. He criticised the fishermen for too late bleeding and gutting the fish (by slitting the throat). In addition, too little salt was used. This led to lower prices and the fact that all Norwegian fish products developed a bad reputation overseas (op.cit.).

After complaints from merchants in Bergen and Kristiansund over poorly processed fish, the natural scientist and vicar P.V. Deinboll (1839) was given an assignment by Selskapet for Norges Vel (a Norwegian welfare company) to write a guide for fishermen outlining the proper processing of clip fish. He pointed out that Newfoundland fishermen gutted the fish at sea while the long lines were being set anew, so that the fish could be salted as soon as possible (op.cit.).

A lack of timely gutting was an ongoing problem. The clip fish was neither gutted nor salted as soon as it was caught, it was not evenly salted, was rinsed too late, was not re-adjusted often enough, and worst of all, it was not pressed well enough (Eidsvaag 1921; Bugge 1923; Moltu 1932). For, when such fish was first stored on land and then transported by ship, the salt was pressed out, which had not occurred earlier, thereby turning the fish raw, wet and later acidic (Op.cit.). Besides, the fish could also be destroyed by halophilic (salt-loving) bacteria or certain strains of mould. This was a result of poor hygiene on-board transport vessels, in the storage houses, and poor salt quality.

Quality could also decline if the fish spent too long in the nets due to bad weather, or if producers purchased poor quality salt (Moltu 1932). Before the time of cold storage, any fish stored throughout the summer would also have deteriorated (Op.cit.). Initially, the development of deep-sea fishing led to a deterioration in quality. Earlier the large smack ships were often out for 3 or 4 days at a time while the fish was kept un-gutted in the hold (Jensen 1963).

All initiatives toward improvement in the 1800s came from private sources, while the public sphere kept a low profile (Deinboll 1839; Eidsvaag 1921; Moltu 1932). Conditions improved after 1900. A private institution was "Selskabet for de Norske Fiskeriers Fremme", a fish promoting company set up in 1879 to modernise the fishing industry. Up until 1898, they sent out research fellow recipients who instructed on proper fish processing and handling. In 1887, the company also initiated a premium fund for dry, handsome, and well-processed clip fish. But when it came to combat clip fish mould the overhead costs became so ponderous that the state had to step in with grant allocations (Selskabet for de Norske Fiskeriers Fremme 1879–1929).

The clip fish industry in general had long been aware that quality issues were related to production and repeated attempts were made to rectify the situation. What, then, are the reasons for which suggestions to improve quality were never followed up?

There appear to be four main reasons for this (Helland 1911; Eidsvaag 1921; Bugge 1923; Moltu 1932). First, fishermen initially complained that they received the same price regardless of quality, and understandably this did little to encourage the acquisition of good products. Some also experienced being paid less for good fish than those who delivered raw and poorly prepared goods, because such fish was heavier (Op.cit.).

Secondly, the large demand from Spain led to the fact that the exporters were less concerned over quality or set prices accordingly. Even if good fish brought a better price in general, the product quality had less of an impact when the demand in general was strong and even poor products could be exported with very good revenues (Helland 1911; Vollan 1956). The large demand owed to the fact that without Norway, production in North America and Europe was too low. Thus, Norway had the double advantage of being able to sell a larger amount of clip fish at evenly increasing prices (Solhaug 1983).

Thirdly, and this had to do with demand pressures, both fishermen and exporters, as Jensen (1963, p. 144) argued, placed greater importance on quantity than quality, a result of the fact that markets for fish, for a large part, were countries where living standards were low. According to the consular reports from Bilbao in the 1880s and 1890s, large-scale exports from Norway resulted in a lot of poor quality and cheap fish on the market (Vollan 1956; Solhaug 1983).

As mentioned, Norway placed great importance on the fact that Spain was a large market of poorly paying consumers. Someone even claimed that inland consumers due to the price put a cheap fish of inferior quality before first-class fish (Johannessen 1928; Jensen 1963; Solhaug 1983).With this point of departure it may seem logical that the Norwegians bet on mass production for the lowest price possible. Still, this view of the market had to collapse the day competitors also competed over price, which occurred in the 1890s (Vollan 1956).

Fourth, there was no quality assurance law. This first came into force in 1921 with *vrakerloven* (the grading law) and in 1933 with the *bløggerloven*, (the law of the timely bleeding of the fish). An independent law regarding gutting has never been introduced. But, what Norway introduced in 1921 was standard-based quality grading, while competitors had accommodated market-based grading (Johannessen 1928). Iceland had already introduced voluntary grading in 1904 after the Spanish had made greater demands of Icelandic fish in 1890s, demands that led to Iceland having to yield compensation for poor fish. Iceland hired five chief grading inspectors and two of these were sent to Spain and Italy to scrutinise the markets. When significant improvements in fish quality due to the voluntary grading were observed, obligatory grading for fish sent to the Mediterranean region was introduced (Vollan 1956).

To explain the downturn in the Spanish market, we must take a retrospective view of the differential trade. But the limitations placed on Norwegian shipping spurred a search for new routes. Moreover, the new wealth and constricted opportunity experienced by entrepreneurs in the Møre region led them to search the world for new markets. They found what they were looking for in South America and the West Indies, markets they had learnt of along the way in their dealings with Spain. Spain operated namely the re-export of clip fish to Latin America (Eidsvaag 1921; Fougner 1941; Vollan 1956). However, the largest amount of clip fish by where by far transported through Hamburg and Altona (now a district of Hamburg), a transport that began in the 1840s. By the 1880s, these cities completely controlled exports from Norway to the American markets. This lasted until World War I when Norwegian overseas liners took over. As middlemen, the German merchants placed strong demands on Norwegian fish while they themselves wished to gain market access. In all practicality, a market grading of Norwegian fish headed for the American markets was undertaken. As a result, Norwegian fish gained a good reputation, but on the other hand, the worst products ended up in Spanish markets (Op.cit.).

3.6 1914–1960: The State Regulations Phase

The outbreak of war in 1914 led to a marked decrease in sales. Simultaneously, the state monopolized the clip fish trade. A central fish processing plant in Ålesund was made responsible for sales from the outset of the 1920s (Verdens Gang 1921), and according to Vollan (1956), Norwegian fish sales experienced serious problems because of poor fish originating from the state-controlled monopoly.

By the end of the 1920s, with the exception of Galicia, the northern Spanish market had been lost. In Bilbao, a city with 100 fishmongers selling clip fish, none of them carried the Norwegian product. The same was true for the wealthy northern Basque provinces and the large inland cities (Waldal 1928; see also Lange 1927).

As had been the case with Catalonia, the Norwegians had not managed to keep up with developments (Clausen 1927; Lange 1927; Johannessen 1928; Møller 1928). That Norway had always been able to unload fish in large quantities, to the point of nearly establishing a monopoly, had given Norwegian producers a feeling that customers needed them more than the other way around. Such attitudes combined with a conservative posture that Norwegian fish was the best no matter what, and declining contact with and sensitivity toward the markets, were obviously destructive for Norwegian exports (Op. cit.).

Consul H.T. Møller (1928) in Barcelona challenged Norwegian exporters to follow the example of others and advertise their fish. One was not only to emphasise the hidden quality, but also to consider appearance and packaging, since Icelandic clip fish was prettier, both whiter and more appetizing (Op.cit.). Icelanders continue to follow the same principle to this day: "What was good for the eye was also good for the stomach" (Lindkvist et al. 2008,p. 115).

Journalist Ø. Lange (1927) admitted that competitors took business-oriented initiatives more quickly and advertised more aggressively. Norwegians, however, had been on the defensive and proud of an old, entrenched position. Furthermore, at the outset of nearly every shipment, agents from the competitors consulted with major importers and undertook market research among consumers to collect information about the quality of the goods, any possible failures, and any changes the customer would like to see. They took immediate note of such competitive sensitive information. As opposed to the Norwegians, rivals also took into account the desires of both importers and consumers (Op.cit.).

They also made personal contact. For example, Icelandic grading agents, exporters and ship-owners travelled to Barcelona in 1921 to gain knowledge of what sorts of demands existed among customers. The Faroe Islands also sent grading agents to Spain and Spanish-speaking German exporters visited most importers in person (Vollan 1956; Fiskets Gang 1914).

Lange (1927) suggests that all Spanish experts celebrated the premium quality and taste advantages of the Norwegian fish. But, it was the customers' perceptions that were decisive and they wanted the snow white Icelandic fish. They were also of the opinion that whiter fish was more resistant to clip fish mould than the yellower Norwegian counterpart. This was purely a fashionable trend; consumers simply did not know what was good for them. In terms of Galicia, this trend had not yet arrived, which he hoped would never happen. However, this did occur the following year.

A major problem was the involvement of Norwegian exporters with direct Spanish domestic sales, which stood in direct competition with their own Spanish customers among the city-based importers. This was the case from the first decade of the 1900s. This ignited anger among Spanish importers, and was undoubtedly an important reason for the intense decline in the sale of Norwegian fish in a city like Bilbao. One importing company that had purchased Norwegian fish for generations switched to the Icelandic version, and they would make sure that such fish would be eaten even in places that had never eaten any other but Norwegian fish (Fiskets Gang 1914; Johannessen 1928). Several small-scale sale firms in Bilbao must also have triggered a campaign against Norwegian clip fish (Thingvold 1928).

The inter-war years were characterised by considerable economic problems and all countries, regardless of political views, approached a subsistence policy with great restrictions on free trade (Hanisch et al. 1999). This was evidenced in Spain's priority of its own trawlers and drying facilities. The new trade policy was based on acquiring the most possible fish without relying on other countries. In accordance with this, a Spanish fishing industry was developed to get direct access to the fishing grounds for cod (Vollan 1956). In 1935 for instance, Spanish trawlers delivered one-quarter of the total import of processed fish (López Losa 2008). It was also in accordance with developments in trade politics that Spain demanded market-sharing agreements in the 1930s. Until then, one had an indirect form of taxation whereby municipalities supported the largest imports possible because this led to toll income for the state. Market-sharing entailed that each country was assigned a clip fish quota that one assumed would be sold on the Spanish market. The quota for Norway was determined in relation to Norwegian purchases of wine and liquor (Vollan 1956; Tande 1957).

Through the trade agreement of 1934, Norway received a prominent clip fish share of 12,800 metric t (Tande 1957). The quota system proved valuable to Norway. While Norwegian exports from 1930 to 1934 made up 27 % of the average yearly Icelandic amount, Norway pulled ahead from 1935 to 1938 (Op.cit.). The main reason was the fact that Norway had more quantity to offer. This is confirmed by developments seen in the first post-war years when Norway performed relatively well on the Spanish market. Bilateral trade agreements favoured Norway, which in opposition to Iceland, bought relatively large amounts of oranges (NOU 1972).

But the civil war in Spain created new obstacles and in 1939, imports were greatly reduced. Norway's quota was cut in half and, in addition, a compensation agreement regarding delivery of clip fish in exchange for various Spanish products was canceled. The Second World War brought nearly all trade to a standstill (Gerhardsen and Gertenbach 1949).

After 1939, an import co-operative took control over the Spanish import (Gerhardsen and Gertenbach 1949; Tande 1957). As a counter act, the Norwegian government entered into a co-operative export monopoly agreement through the national Norwegian clip fish exporters' union. The sold quantities were divided between participating exporters according to quotas. No longer did anyone have the legal opportunity to negotiate with or sell to Spain apart from through the co-operative. (Tande 1957). From 1945 to 1952, the Norwegian clip fish and salted fish export council was responsible for sales, after which it appointed A/L Unidos to do the job. In other words, centralised export and import systems were born (Op.cit.).

The tough restrictions lasted until liberalization of the Spanish economic policy in 1959, among other things, through relief from certain import restrictions. Norway ended up meeting its old Spanish clip fish customers as competitors, both at the fishing grounds and on the clip fish market (Tande 1957, Gerhardsen 1964).

After World War II, Spain state support built up a modern Spanish trawler fleet with a significant capacity to meet demands. From the mid-1950s, the catches increased greatly along the Newfoundland fishing banks, but much of this was small-sized fish (Tande 1957). Denmark and the Faroe Islands were the largest purveyors of clip fish to the Spanish market during the first post-war years, followed thereafter by Norway. Iceland made a comeback in 1952 and had already become dominant compared to Norway, while Newfoundland played an insignificant role (Tande 1957).

Centralised export to Spain in the 1940s and 1950s had a strong influence on quality. Since Spain consisted of a range of different markets that demanded different requirements of their fish products, one had to take into consideration that certain fish were destined for certain districts. A development of a classification system differentiated the products regarding the degrees of dehydration, size and quality supported these considerations (Tande 1957).

But, according to Tande (1957), individual exporters had no direct contact with the markets and had gradually lost their knowledge of the particular demands posed by the customer. He maintains, that under free trade, individual export firms had specialised sales to certain districts and thus became familiar with customers' needs through personal contact and through that of agents. In the 1940s and 1950s, however, all firms had to supply a portion of the assigned fish size according to the quota. Such sales were of the same size and quality and of the same degrees of dehydration as was specified in the contracts as a whole. Tande (1957, p. 376) argued that, in practice, it was difficult for exporters and the Norwegian Grading Department to reach agreement. Spanish buyers were dissatisfied with the quality, and raised claims to considerable recompense. The Norwegian representatives in Spain had to admit that these shipments were of poor quality.

3.7 1960–2011: Towards a Freer Trade

In the 1950s, Spain lived through an economic crisis with large deficits in the balance of trade and rapidly rising inflation. However, 1957 was preface to an economic change of course. In 1959 a stabilisation plan came about (Tamames 1974). The Organization for European Economic Co-operation (from 1961 the Organization for Economic Co-operation and Development) and the United States of America gave sizeable loans and cash gifts in exchange for Spain fulfilling certain stipulations. The most vital stipulations included relief of restrictions towards imports and foreign investment. A rapid industrial and economic growth took root as many left towns and villages in the countryside and took jobs in industrial and service sectors. The economy was further liberalised after the fall of the Franco regime in 1975 and membership of the European Union in 1986 (op.cit.).

In the decade between 1960 and 1970, Spain was the largest producer of salted fish in the Atlantic territory with a yearly average of 87,000 metric t. But in the early 1960s, demand for small-sized fish fell dramatically. This development occurred at the same time as, and was likely a partial result of, a strong increase in private purchasing power and the introduction of frozen fish to the market. At the same time, domestic demand fell to around 50,000 metric t. This pressured the Spanish into the export market. Clip fish of small-size went to low-paying markets like northern Brazil and the Congo, while salted fish was sent to Portugal and Italy. In the same period, a certain shift of domestic marketing occurred, and in 1970, 20–25 % of consumption was sold pre-packaged (NOU 1972). Still, Spain had to import some qualities of salted fish that the country could not provide sufficient quantities of themself. Some smaller quantities of the imported fish was also re-exported (Op.cit.).

Until 1960, Norway had a bilateral trade agreement with Spain, but with the transition to a multilateral system, Norwegian export quotas were no longer "protected", and the entire Norwegian clip fish export system collapsed (Malmo 1964). From 1968 to 1970, there were no Norwegian clip fish exports to Spain (NOU 1972).

Wet salted cod was not a common product in Spain before the introduction of cooling technology. Before 1970 the Spanish import of wet salted cod was minimal, but this changed due to a decline in Spanish production in the 1970s. The Norwegian export to Spain was stable in the 1970s and 1980s with a yearly average of 2000 metric t of salted fish. Some of it was clip fish, but it became increasingly marginal compared to wet salted fish (NOU 1972; Fiskets Gang 1980).

In the 1990s the Norwegian export rose to 14,000 metric t due to the collapse of the Newfoundland fisheries and a decline in the Icelandic fisheries. The market share reached 34 % in 1998. But when the Icelandic fisheries recovered, the Norwegian share declined from 1999 and is down to 10 % in the 2000s. Today Norway exports 6000–8000 metric tonnes of salted fish to Spain (Lindkvist et al. 2008).

3.8 Conclusion

This chapter has laid out the most important structural and institutional constraints for the development of value chain conventions and performance in the Norwegian-Spanish salted fish trade since 1665. Norway has entered a path with a focus on cost-oriented production conventions in the initial parts of the value chain quantity to get provision of the country's large fish resources. This is in contrast to the more market-oriented convention practiced in Iceland, which has access to the same kind of fish species as Norway but in a lesser quantity. Iceland has taken advantage by offering the right quality for the best paying end market by responding to consumer conventions and preferences, while Norway has been exporting a low-cost standardized salt fish for low-income markets without with making an effort to respond to the quality claims from the market. Icelanders had clearly invested in market-oriented activities since the 1890s, taking Spanish complaints into consideration in their product offerings and accommodating the market's fluctuating wishes and needs. Meanwhile Norway, as a cost-oriented volume producer-oriented exporter, has demonstrated little interest in marketing investments to take advantage of the differences in the market's quality preferences by learning from its mistakes and adjusting its product qualities; thus it has continued on the same familiar track. Ironically, Norway has reaped large rewards from state intervention and regulation in the form of differential tolls, quota trade and bilateral agreement, while Norwegian exporters have experienced substantial hardship under more liberal trade conditions.

References

Berg H (1927) Klippfiskeksporten fra Trondhjem. Miscellanea Nidrosiensia. Trondhjems historiske forening, Trondhjem

Berg T, Hanisch T, Lange E, Pharo H (1988) Norge fra U-land til I-land, 1830–1980, 2nd edn. Vekst og utviklingslinjer. Gyldendal, Oslo, 1830–1980

Blich JW (1928) Norges klippfiskeksport i tiden 1890–1925 statistisk belyst Årsberetning vedkommende Norges Fiskerier 1927—Nr. X. Fiskeridirektøren, Bergen

Bugge K (1923) Aalesunds historie. Volume I. Aalesund kommune, Ålesund (Published by: in connection with the 75 years jubilee of the town. MCMXXIII)

Clausen don P (1927). Klipfiskhandelen i Spanien de siste 50–60 år. Norges Handels- og sjøfartstidende 30.8, Oslo

Deinboll PV (1839) Om Behandlingen og Tilvirkningen af Saltvandsfisk. En Veiledning for den fiskende Almue i Norge. Den Kongelige Norske Regjerings Finants-. Handels- og Told-Departement, Christiania

Dyrvik S, Fossen AB, Grønlie T, Hovland E, Nordvik H, Tveite S. (1979) Norsk økonomisk historie 1500–1850. Universitetsforlaget, Oslo

Eidsvaag E (1921) Om tvungen vrakning av klipfisk. Norsk Fiskeritidende, Bergen, pp 50–54

Fiskets Gang (1911) News article: September 27. The Norwegian Directorate of Fisheries, Bergen, pp 267–270

Fiskets Gang (1914) News report. No 10. The Norwegian Directorate of Fisheries, Bergen, pp 93–94

Fiskets Gang (1980) News reports. No. 3, The Norwegian Directorate of Fisheries. Bergen

Fougner S (1941) Norsk eksport av tørrfisk og klippfisk. Cand.philol. dissertation in Geography. Universitetet i Oslo

Gerhardsen GM (1964) Fiskeriene i Norge. Økonomi og politikk. Universitetsforlaget, Oslo

Gerhardsen GM, Gertenbach LPD (1949) Salted cod and related species. FAO Fisheries Study No. 1. Food and Agriculture Organiszation of the United Nations

Hanisch TJ, Søilen E, Ecklund G (1999) Norsk økonomisk politick I det 20. århundre. Høyskole-forlaget – Nordic Academic Press, Kristiansand S

Helland A (1911) Norges Land og Folk. XV. Romsdals Amt. Volume I. Aschehoug, Kristiania

Historisk Statistikk (1968) Norges offisielle statistikk XII 245. Statistisk Sentralbyrå, Oslo 1969

Innis HA (1954) The cod fisheries. The history of an international economy, 2nd edn. University of Toronto Press, Toronto

Jensen M (1963) Norges historie. Unionstiden 1814–1905. Universitetsforlaget, Oslo

Johannessen B (consul-general in Bilbao) (1928).Klipfiskmarkedet i Spanien. Bergens Tidende 8.8, Bergen

Johnsen AO (1942). Kristiansunds historie. Volume I. First half-binding, Kristiansund
Lange Ø (1927) Norsk klipfisk i Spanien. Norges Handels- og Sjøfartstidende 12.8, Oslo
Lindkvist KB, Hauge ES (2000) Innovasjon i norsk saltfiskindustri i 1990-årene. Noen eksempler. Serie A: Meddelelser fra Institutt for geografi, Nr. 240, University of Bergen, Bergen
Lindkvist KB, Gallart-Jornet L, Stabell MC (2008) The restructuring of the Spanish salted fish market. Can Geogr 52:1
López E (1999). Technical change in the basque fisheries: the diffusion of steam trawling. Studia Atlantica, 3. Fiskeri- og søfartsmuseet, Esbjerg, pp 225–252
López Losa E (2008) Spanish Atlantic cod (Gadus Morhua) fisheries in Newfoundland in the second half of the twentieth century. Can Geogr 52:1
Malmo E (1964) Eksport til Spania. Håndbok for eksportører. Norges Eksportråds markedshåndbøker. Serie 111. Nr. 10. Norges Eksportråd, Oslo
Moltu P (1932) Fiskarsoga frå Sunnmøre og Romsdal. Sunnmøre og Romsdals Fiskarlag, Aalesund
Møller HT (consul-general in Barcelona) (1928) Mister vi vært marknad i Spania? Tidens Tegn 5.11, Oslo
Norges Handels- og Sjøfartstidende (1928) News article 17.8, Oslo
NOU Klippfisk- og saltfiskeksporten. (1972):26
Pedersen T (1992) Prosesser og produkter i norsk fiskeindustri. Bind 2. 2nd edn. Universitetsforlaget, Oslo
Piquero S, López E (2006) New evidence for the price of cod in Spain: the Basque Country. Studia Atlantica, 8. North Atlantic Fisheries History Association, Maritime Historical Studies Centre, University of Hull, Hull, pp 195–211
Puerta Rueda N de la (1998) The codfish trade in the Bay of Biscay in the second half of the nineteenth century. Studia Atlantica, 9. Fiskeri- og Søfartsmuseet, Esbjerg, pp 191–202
Rathke J (1907) Afhandling om de norske Fiskerier og Beretninger om Reiser i Aarene 1795–1802 for at studere Fiskeriforhold m. v. Selskabet for de norske fiskeriers fremme, Bergen
Selskabet for de Norske Fiskeriers Fremme (1879–1929)
Solhaug T (1983) De norske fiskeriers historie 1815–1880, 2nd edn. Universitetsforlaget, Oslo
Strøm J (ed.) (1949) Norsk fiskeri og fangst håndbok. Vol. 1. Alb. Cammermeyers Forlag, Oslo
Tamames R (1974) Estructura económica de España. Guadiana de Publicaciones, Madrid
Tande T (1957) Norsk fiskeripolitikk. En analyse av fiskerinæringens utvikling siden 1920. Studieselskapet samfunn og næringsliv, Oslo
Thingvold A (fisheries agent in Spain) (1928) Fiskeriagenten i Spanien klager over propagandaen mot norskfisken. Aalesunds Avis 20.12, Ålesund
Verdens Gang (1921) News article, 4.3, Oslo
Vollan O (1956) Den norske klippfiskhandels historie. Øens Forlag, Førde
Waldal S (1928) Den norske klipfisk utkonkurreres mere og mere i Spanien. Morgenbladet 12.2, Oslo
Wallem FM (1893) Handelen med Tørfisk og Klipfisk. W.C. Fabricius & Sønner, Kristiania
Wedel Jarlsberg F (1932) Reisen gjennom livet. Gyldendal, Oslo

Chapter 4
Explaining the Mismatch Between the Norwegian Salted Cod Value Chain and the Spanish Fish Market

José Luis Sánchez-Hernández

Abstract Spain has a long tradition of salted fish consumption. Norwegian salted cod enjoyed a good reputation in Spain for its regular supply and traditional taste. Nevertheless, between 1978 and 1986, Spain lost access to cod fishing grounds in the North Atlantic and importation turned out to be the only way to feed Europe's second largest per capita consumption market for salted fish products. Norway became its main cod supplier until the late 1990s. However, the Norwegian salted cod industry has failed to keep this strategic market because its leading actors have not renegotiated a new definition of quality attuned to the current patterns of fish consumption in Spain.

Keywords Fresh fish · Quota system · Value chain · Consumption trends · Mediterranean diet · Food origin

4.1 Introduction

Food value chains (FVCs) rest upon a basic and shared agreement about two main points (see Chap. 2): how to efficiently arrange the production system and how to determine attributes that consumers expect to find embedded in its core foodstuff. When FVCs' actors agree upon both issues, the competitive success of the whole industry is the most likely outcome. Actually, consumers in developed countries are putting pressure upon FVCs worldwide to reshape those agreements because new market niches are emerging where criteria other than price and utility influence householders' purchasing behaviour. These changes in consumers' preferences mean that new rules must be negotiated all along FVCs to face competition in markets that are no longer homogeneous, but increasingly differentiated.

Nevertheless, mis-coordination may arise when any given FVC's long-term agreements are challenged by the ongoing qualitative changes in food markets. This chapter argues that, if the most powerful FVC partners are reluctant to update their practices and to attach new attributes to their core foodstuff according to emerging societal discourses (Ponte and Gibbon 2005) about quality, a loss of market share may be

J. L. Sánchez-Hernández (✉)
Department of Geography, University of Salamanca, Salamanca, Spain
e-mail: jlsh@usal.es

K. B. Lindkvist, T. Trondsen (eds.), *Nordic-Iberian Cod Value Chains*,
MARE Publication Series 8, DOI 10.1007/978-3-319-16405-2_4

expected which threatens the coherence of the trade flow from rural producers to urban consumers.

The performance of the Norwegian salted cod industry in the Spanish fish market for the last decade provides empirical evidence for this argument[1]. Several chapters in this volume carefully measure this process. In short, the argument assumes that Spain has a long tradition of salted fish consumption because meat intake is banned during Catholic Lent. In this culturally shaped dietary context, Norwegian salted cod enjoyed a pretty good reputation for its regular supply and traditional taste, mostly appreciated in rural and sparsely populated areas where other fish species were hardly available (Gago 2007). Nonetheless, after an enlargement of the marine exclusive economic zones to 200 nautical miles in 1978, Spanish trawlers were pushed away from some traditional fishing grounds in the western Atlantic Ocean (Newfoundland). Later on, Spain was forced to accept a big reduction of its fishing quotas and catching capacity as a prerequisite to become a member of the European Union in 1986. Importation became thus the only way to feed Europe's second largest per capita consumption market for salted fish products, just behind Portugal (NSEC 2006). Norway took advantage of its traditional trade links with Spain (Richter-Hanssen, this volume) and became its main cod supplier, holding one-third of the salted fish market in the middle of the 1990s (Lindkvist et al. 2008).

Despite this primacy, the Norwegian salted cod industry has failed to keep this strategic market. The Norwegian contribution to Spanish codfish weight imports fell from 7.5 to 4.3 % between 2000 and 2007. Price figures are also meaningful: for a Spanish mean import value per kilo of 4,330 € in 2000 and 4,520 € in 2007, Norway underscores with 4,160 and 4,400 €. These data suggest that Norwegian cod sales to Spain remain attached to low value-added products, mostly the traditional dried and salted cod (*clipfish*), which meets increasing problems to compete with more convenient cod products delivered by Iceland, the new major player in the Spanish cod market (Xie and Myrland 2010).

Therefore, the Norwegian salted cod FVC is losing positions because its leading actors have not renegotiated a new definition of quality attuned to the current patterns of fish consumption in Spain. Norwegian companies and authorities are simultaneously failing in the communication of their strongest values to Spanish consumers and in adoption of the attributes demanded by them. The discussion of this argument will take four steps. First, the main concepts are outlined with reference to the theoretical framework developed in Chap. 2. Second, the overall Spanish fish market is analyzed in terms of industrial, market, domestic, public, civic, and ecological

[1] The theoretical framework has been developed with the support of the research grant "Governance, conventions and innovation in the wine districts of Castile and Leon: typology and prospective of Designations of Origin based on the theory of worlds of production", funded by the Spanish Ministry of Science and Innovation (ref. CSO2008–05793-C03–01/GEOG) and the Regional Government of Castile and Leon (ref. SA080A08) for the period 2008–2011. The salted-fish case study is based on the research project "The Spanish salted-fish market and opportunities for the Norwegians" (2008–2011), funded by the Norwegian Research Council (ref. 185126/I10).

conventions. Third, the analysis of the Norwegian salted cod FVC uncovers the mismatch between its core practices and the prevailing values in Spain: disagreement between producers and consumers becomes thus the key to understand that decreasing market share. The final section sums up the findings and discusses opportunities to overcome this situation which has profound implications for regional development in northern Norway.

Government reports, statistical databases and personal fieldwork support the survey on the Spanish seafood market. The salted cod section is indebted to similar sources, but it mostly relies on the large repository of qualitative and quantitative information available for partners of the research project that frames this book (see footnote #1). It includes in-depth interviews with Norwegian and Spanish players, namely fishermen's representatives, processing companies, foreign traders, research and marketing organizations, wholesalers, retailers, consumers, and restaurants.

4.2 Orders of Worth and Food Value Chains

Neoclassical economic theory acknowledges price and utility maximization as main criteria for producers and consumers to take their decisions in the marketplace. Nevertheless, and focusing on the firm as a field of discussion, Boltanski and Thévenot (2006/1991) demonstrate that even a profit-seeking organization calls upon more principles or *orders of worth* than just price and utility in the process of qualification, evaluation or *justification* of its own rules, behaviours, decisions and products. In addition to *market* or *commercial* qualities (products must be affordable, useful and profitable), any industrial product -say food in this context- is often assessed in terms of the standardization and technical efficiency of its production process (*industrial* qualities), of its geographical or personal origin (*domestic* qualities), of the reputation of its trademarks and branded producers (*public* qualities), of its contribution to community development, social justice and public health (*civic* qualities), and of its implications for environmental sustainability (*ecological* qualities) (Murdoch et al. 2000; Morgan et al. 2006; Lindkvist and Sánchez 2008; Sánchez et al. 2010). This sixfold typology may easily be re-arranged around three core questions: *what* is this product? (Qualification by industrial and market qualities); *who* is the supplier? (Qualification by domestic and public qualities); and *how* was it made? (Qualification by civic and ecological qualities).

Therefore, quality might be defined as the particular combination of market, industrial, domestic, public, civic, and ecological attributes for food that complies with the expectations of all FVCs' actors and respects the agreements achieved: "*Quality* [. . .] *is a datum constructed by producers and consumers; it is then known, recognized, and incorporated into the expectations of both parties*" (Storper and Salais 1997, p. 38). Section 2 draws on this strand of literature to display the main quality preferences of Spanish seafood consumers.

Accordingly, to meet demands from market niches framed by more complex combinations of orders of worth, industrial companies must continuously restructure their basic conventions, or in Storper and Salais' (1997) conceptualization, the primary rules and decisions needed for every industrial market to be settled and sustained over time. Both authors list the following conventions: what goods to produce (generic-consolidated for homogeneous markets, or dedicated-customized for specific groups of customers); what resources to employ (specialized-localized in a place or specialists' community, or standardized-ubiquitous and available in many locations); which actors are allowed to join the production system (low entry barriers and participation open to newcomers, or complex entry barriers which imply restriction to members of a community); how to define actors' identity (abstract-interchangeable because quantity is the basis of the value chain, or personalized-unique because skills and differentiation are key for competition), and what role should be assumed by the state (absent and only setting the basic conditions for competition; external and just supplying resources to private players; or situated in networking with all stakeholders for a joint achievement of shared decisions and goals). Section 3 resorts to these categories to explain why the Norwegian salted cod FVC is failing to follow the qualities demanded by the Spanish seafood market.

4.3 Orders of Worth and Food Qualities in the Spanish Seafood Market

Spain consumed 1,580,170 metric t of seafood in 2009, a figure which includes fresh fish, frozen fish, salted, dried, smoked and canned fish, every type of shellfish, and farmed fish as well. National vessels landed two thirds of that amount, the rest being imported from France, the United Kingdom, Portugal, the Netherlands, Morocco, Argentina, China and Ecuador as main suppliers (MERCASA 2011). As Larsen (this volume) notes, Spain is the largest seafood consuming country in Europe by volume and the second per capita, with 33.8 kg by person in 2009, an average which has increased from the 29.7 kg recorded in 1997. The share of seafood in the Spanish expenditure on food for home consumption is 13.5 % (Ministerio de Medio Ambiente y Medio Rural y Marino 2009). Within this broad and encouraging context, codfish accounted for 2.7 % of seafood consumption, both by volume and expenditure, in Spanish households in 2009 (Household Food Consumption Database online). In the same year, Norway exported 18,850 metric t of seafood to Spain, 6,965 of which fell into the different categories of cod recorded by official statistics (DATACOMEX online) (Fig. 4.1).

According to all sources reviewed, the most important *commercial qualities* in the Spanish seafood market are freshness and whiteness. Two thirds of all the seafood caught by the Spanish fleet (0.79 million t) is landed fresh. About 53 % of Spanish households are accustomed to buying fresh seafood, despite its comparatively higher prices. Actually, only 13 % of customers report that price is a key factor in their purchase decisions regarding seafood, because price is often valued as a proxy for quality in the fresh seafood market segment. Accordingly, fresh fish purchasing frequency is 7.6 days a month, Mondays being usually avoided by experienced

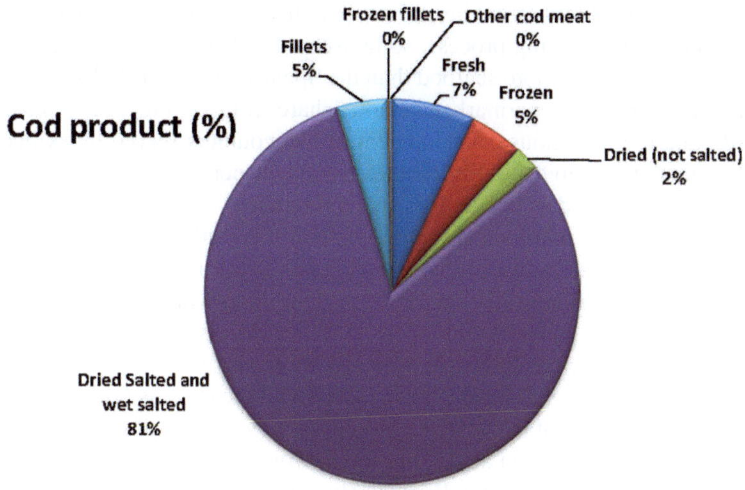

Fig. 4.1 Cod imports from Norway to Spain in 2009. (Source: author's elaboration with information from DATACOMEX and Seafood Norway [online])

householders who avoid buying fish stored over the weekend in retailers' freezers. About 61 % of seafood volume was sold fresh in 2009 and white species like hake, sole, bream, bass, monkfish, or turbot, are mostly appreciated by consumers due to their natural taste and juicy texture (Ministerio de Agricultura, Pesca y Alimentación 2007; MERCASA 2011).

A lower 42 % of Spanish households report usually buying frozen seafood, which is mostly stored at home as a supplement for unexpected events or as an ingredient for sauces, puddings or other recipes whose taste does not fully depend on freshness. In addition to the traditional slaughtering of fish upon customers' request at the fishmongers' counter, processing companies are launching new products like fresh fillets and packaged loins, more convenient for singles and couples, and mostly sold at supermarkets. Nevertheless, according to consumer surveys (Ministerio de Agricultura, Pesca y Alimentación 2007), processed fish items are valued as lower quality than pure fresh seafood because the true colour, shape, and thickness of the content cannot be assessed before purchase: more than half of the population either never buy them or is not aware of its availability (Ministerio de Agricultura, Pesca y Alimentación 2008). These quality demands appear to be so deeply rooted in Spain that the immigrant population has adopted the same consumption patterns as the local population, ranking hake (*merluza*, the hallmark for fresh fish, with 25 % of sales in that segment) at the top of their preferences (Ministerio de Medio Ambiente y Medio Rural y Marino 2009, Household Food Consumption Database online).

Freshness and whiteness are thus cornerstone attributes of the Spanish fish market. Both of them strongly shape the *domestic qualities* attached to the process of selling and buying fish. Traditional fishmongers still keep a market share of 39.4 % in 2009 (MERCASA 2011), which increases to 49.3 % in the fresh fish market, because only reliable sellers can guarantee freshness. Those figures clearly show that trust is involved when valuable food items like fresh fish are traded. Of course, this is a time-consuming purchasing process, so households where one adult member does not hold a job consume more seafood than the Spanish average (MERCASA 2011). In addition, the rise of supermarkets' market share from 28.8 to 41.3 % since 2000 is mostly due to the in-house development of fish counters served by skilled staff, a strategy that strives to capitalize on the advantages perceived by consumers who usually buy at independent specialized seafood retailers.

This interpersonal relationship between small fish outlets, either neighbouring shops or stalls in larger marketplaces and their usual customers, is also underpinned by other domestic attributes like service and counselling which, perhaps, justify the higher average prices paid at fishmongers' facilities (Ministerio de Medio Ambiente y Medio Rural y Marino 2009). At least 40 % of customers ask for information about fish origin (another domestic trait), 20 % about the process (either fresh or de-frozen) and 9 % about suitable recipes. Moreover, 55 % of consumers who complain about the lack of information at the sales point would like to know more about fish origin (Ministerio de Agricultura, Pesca y Alimentación 2007). Formal certification schemes like traceability have not proved their reliability yet, since consumers in discussion groups arranged by Spanish food authorities report little or no knowledge about their meaning or implications, which is consistent with the findings of Eden et al. (2008) in the United Kingdom. Regarding this point, restaurants must actually be mentioned as key nodes for attaching domestic qualities to fish species served in their menus and recipes. The influence of chefs and TV cooking programs (Larsen, this volume) also increases domestic values because they often link fish consumption to the fashionable and locally embedded Mediterranean diet: 20 % of consumers report that their commitment to these habits is a main reason for eating fish (Ministerio de Agricultura, Pesca y Alimentación 2007; 2008). Therefore, seafood, and namely fresh fish, is widely available in Spain.

According to the Household Food Consumption Database online, urban households consume a bit more seafood than rural ones: in settlements under 2,000 inhabitants, per capita consumption at home is only 5 % below the Spanish average. Such a widespread availability rests upon additional *industrial qualities* that consumers usually take for granted. Fish is very often served both in households (ten times a month, Ministerio de Agricultura, Pesca y Alimentación 2007) and in the HORECA channel. Companies and consumers are used to find a variegated fish assortment in wholesalers and retail outlets because the Spanish seafood FVC is able to supply raw material across a large national market in due time and quantity.

There is no significant difference in fish consumption between inland and coastal regions: strikingly, the Canary and Balearic islands are the less fish demanding regions in Spain (17.7 and 16.5 kg by person in 2009 respectively) whereas Castile and Leon, far away from the seashore, ranks secondly with 38.6 kg (Fig. 4.2). By the way, salted and dried cod used to be the main fish specialty consumed in inland Spain until the 1960s (Gago 2007; Richter-Hanssen this volume). Super- and hypermarket chains have developed very efficient logistic systems to source their shops at low prices and they are leaders in the frozen seafood market. However, their corporate skills are supported by a complex network of small and medium actors that includes

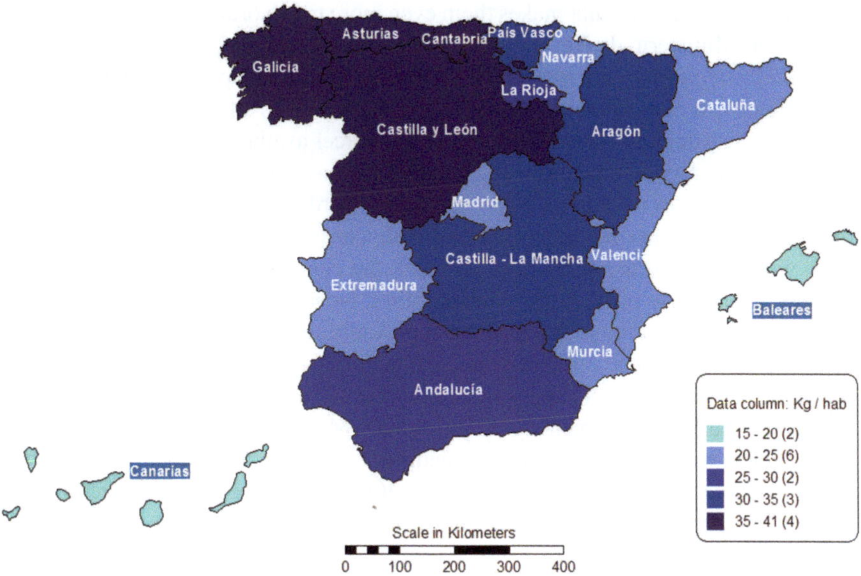

Fig. 4.2 Seafood consumption in Spanish regions, 2009 (kg per person). (Source: author's elaboration with data from Home Food Consumption Database [online])

landing ports, aquaculture facilities, importers, auction sales (*lonjas*), processors, wholesalers (*mercas*), and traders which also supply every sort of seafood to retailers, fishmongers, the HORECA market, and other exporters to third-party countries.

This efficiency of the Spanish seafood value chain is probably influencing the development of the *public* and the *civic* orders of worth. Canned, frozen, filleted and ready-to-eat/ready-to-cook seafood products are usually branded by a number of well-known companies (*Calvo, Isabel, Pescanova, Ortiz, Findus*...). Spanish consumers' commitment with a public attribute like brands, a trait of the whole Spanish food market, is also apparent in this case (Ministerio de Agricultura, Pesca y Alimentación 2007). Fish is also commonly linked to a key civic quality: health. An increasing number of customers (69 % in 2003 and 76 % in 2007) believe that eating fish is a healthy practice; an additional 32.5 % prefer fish to meat, while 64 % enjoy its taste (Ministerio de Agricultura, Pesca y Alimentación 2008). These preferences are reinforcing the pervasive dominance of freshness because 53 % of consumers believe that processed fish products are less healthy than fresh ones (Ministerio de Agricultura, Pesca y Alimentación 2008). Therefore, innovative products like light salted and de-salted fish fillets may benefit from developing public and civic attributes: producers' brand name and health care by means of fish consumption and low salt intake. Regarding the influential role of civic attributes, it must be said that no food scares involving fish have ever happened in Spain, which is consistent with the reliance on FVC players' fairness reported by separate focus groups of traders and consumers. It is also noteworthy that fresh fish dealers are found to be the most concerned in the Spanish seafood value chain when complying with the new rules about food safety, a fact that makes them even more trustworthy from the consumers' viewpoint (Ministerio de Agricultura, Pesca y Alimentación 2007).

Ecological qualities are increasing their influence in the Spanish fish market. The mainstream discourse about fish-farming (or aquaculture) addresses its contribution to mitigate fish depletion and to keeping sea resources available for future generations and for fishing communities as well. Aquaculture has undoubtedly turned expensive species (salmon, trout, bream, bass) into very affordable ones, thus contributing to the general preference for fresh fish. Available consumption data do not tell farmed from wild fish, so it is not possible to check the contribution of aquaculture to the aforementioned rise of fresh fish consumption. Nevertheless, it is quite reasonable to make an argument for such a relationship. Additional support for this claim stems from the fact that more concerned and affluent consumers have started to ask for information about the fish supply system and usually prefer the wild to the farmed (Ministerio de Medio Ambiente y Medio Rural y Marino 2007). Awareness of growing criticism against health-care and feeding systems in fish farms on the part of these consumers may lead to a sort of market divide between fresh farmed fish for low-income people and fresh wild fish for the wealthier and for the upper segment of the HORECA channel.

4.4 Conventions and Practices in the Norwegian Salted Cod Value Chain

4.4.1 Protecting the Sea: Restricted Participation and Abstract Identity

Resources from the sea are essential in the Norwegian economy and welfare. Fishing and aquaculture accounted for 0.5 % of Norwegian gross domestic product in 2009. About 960,000 t of fish (mostly salmon) were farmed and, 2.53 million metric t of wild fish were landed by 12,730 fishermen in Norwegian harbours in 2009 (Statistics Norway 2009, Statistics Norway [Online]). Sea affairs are at the forefront of Norwegian regulation. Fishermen's historical contribution to the country's economy and identity is reflected in a strong commitment to the preservation of coastal villages and communities (Cruickshank et al. 2009). Sustainability is also embodied in legal arrangements that attempt to promote an environmentally friendly fishing economy (Kristoffersen and Young 2010).

Since 2000, the Law of Participation only allows fishers, with their respective vessels, to be entitled with fishing quotas. Raw material procurement is supposed to be a business for small independent boats that catch, land and sell the fish to factories or traders. The purpose of the Law of Participation was to prevent fishermen from becoming employees for larger vertically-integrated firms, and tried to keep their independence and wealth within self-sustaining coastal communities. The current situation is that fishermen frequently are employed on trawlers owned by powerful shipping companies, which unload the catch in the major cities.

Sustainable management of fisheries is the second main goal of Norwegian regulation. Careful monitoring of fish stock and subsequent decisions on yearly quotas try to adapt the catching capacity to variable resources. Due to its deep social, economic and geographical implications, this complex decision-taking process constitutes a central duty of national authorities, namely the Norwegian Ministry of Fisheries and Coastal Affairs. This process often involves multilateral diplomatic talks about fishing rights with powerful neighbours and competitors like Russia, Iceland and the European Union.

Such a political commitment with these civic and ecological values shapes industrial practices as well. Regulations on seafood safety strictly forbid the use of any chemical additives in Norwegian brining solutions. This way, the salted cod keeps its traditional golden colour, a natural trait of coastal cod, whose muscles become thinner, stronger, and quite darker after swimming against marine streams in its seasonal migration from the Barents Sea southwards to the Lofoten Islands. Icelandic factories, for instance, use to add polyphosphates to the brine mixture since the 1990s to get a whiter product (Lindkvist et al. 2008). Their thick and white cod loins look pretty much like some aforementioned fish species (hake, sole, bass, monkfish, turbot) which Spanish consumers usually buy fresh and that are widely appreciated as tastier, healthier, and higher-quality.

4.4.2 An Engaged (but Misguided) Seafood Innovation System

The economic and cultural meaning of fisheries, and the Scandinavian convention of an engaged state closely connected to societal concerns (Kristoffersen and Young 2010) explain the constitution of a tight institutional network around the fish value chain in Norway. The coalition between the specific ministry for marine and coastal affairs and the influential fishermen associations mobilizes a huge amount of financial and knowledge resources. A 0.3 % levy on all seafood exportations is charged to fund applied research by a wide range of scientific organizations. An additional 0.75 %is also charged for fuelling the Norwegian Seafood Exportation Council's budget, which covers a wide scope of foreign market promotional initiatives.

These research efforts also underpin the conventions of restricted participation and abstract identity which are intended to preserve sea resources. Sustainable management of fish stocks, setting of allowable quotas, aquaculture as an alternative to resource depletion, health effects of fish intake, traceability along the value chain... receive most of funding allocation and are highlighted in all the marketing devices delivered by the aforementioned organizations. Data from NMFCA (2010) about the distribution of R&D expenditure in the marine sector in 2007 display that only 5 % was allocated to market and organisational issues and an additional 6 % to the processing industry (including quality control). Aquaculture received a substantial 27 %, and the rest was devoted to nature-related issues like marine biodiversity, stock surveillance, oceanography, bio-technology, bio-economics and bio-statistics.

This sea-laden bias of the fish innovation system encloses a double contradiction. First, Lindkvist (2010, p. 11) contends that the allocation of many resources to topics that reinforce the ongoing misleading situation creates the mirage of a highly dynamic research environment that actually supports the status-quo. For instance, the strong focus on seafood safety implicitly hides the more valuable civic and natural qualities of seafood because it drives consumers' attention to the standardized technologies used throughout the industrial process undergone by fish from sea to table, whereas most of Spanish customers have not ever heard of the very concept of traceability (see Sect. 2).

In addition, the current foreign market campaigns do not persuade final consumers about these civic and sustainable qualities hidden in Norwegian salted cod production. Domestic qualities like tradition, craftsmanship, and geographical origin are very important in Spain as well. Nevertheless, they are hardly addressed in the too generic Norwegian research and promotion guidelines[2]. Therefore, Norway's worldwide reputation as a high-quality seafood producer (e.g. salmon) remains an undercapitalized asset. Whereas Mediterranean countries are very sensitive to geographical origin as a distinctive attribute of food (Parrott et al. 2002; Ilbery et al. 2005; Sánchez et al. 2010), current claims on cod traceability as a warranty for health are not connected to any particular place and invoke either abstract identity or industrial procedures instead of any domestic attributes.

These two contradictions (the focus on standardized technologies and the orientation to generic markets) point to the core weakness of the Norwegian foreign market strategy: the lack of a single and understandable discourse of quality for salted cod, followed by a coherent set of actions emphasizing both the sustainability of the fishing systems and the production process (a more specialized convention of technology), and the domestic and civic attributes of the product and its suitability for non-mainstream retail outlets (a more dedicated market convention). A number of features (coastal environment, fishing communities, food safety, cod traceability, healthy diet) and market niches (from hard discount outlets to gourmet shops) are actually addressed by the salted cod FVC, but a unifying narrative is still to be constructed and transmitted for potential consumers.

4.4.3 Standardized Technologies for Feeding Generic Markets

Stakeholders across different stages are concentrated in their own tasks and have little influence on the overall development of the salted cod FVC. Independent fishermen remain committed to their traditional technical conventions regarding catches schedule, catching methods, and handling procedures. They hold the quotas and have

[2] Seafood Norway's Spanish branch office launched in 2010 a marketing campaign focused on restaurants. The distinctive qualities of 'Traditional Norwegian Cod' are highlighted in a promotional video clip, labels and brochures. This recent shift implicitly acknowledges that such a topic has been quite neglected up to date.

the right to sell each landing as a single batch, with no grading either of the species caught or of their suitability for different consumers' demands: restricted participation, abstract identity and generic market orientation are thus interwoven. Hence, regional fishermen trade unions are entitled with the monopoly for selling fresh fish and set the minimum prices to be paid in auctions (Lindkvist 2001; Trondsen (b) this volume). These two facts sustain a self-reinforcing behaviour that attaches fishermen to bulk market conventions and makes them unaware of emerging trends in highly-differentiated consumer markets like Spain.

Institutional barriers to the vertical integration of the value chain explain why industrial companies remain so concentrated on resource procurement to meet their contracts with Spanish importers. These agreements are long-term agreements based on carefully detailed standards to match the specific qualities demanded by the Spanish market. This mis-coordination implies that the lack of control over raw material is highly problematic to many Norwegian processors. The heavy burden of the fishing realm makes innovation both more difficult and less profitable: few product innovations have been recorded in the Norwegian salted cod industry for recent years (Lindkvist 2010). Cod is still the most valuable wild seafood Norwegian exportation, but two thirds of that value still comes from the most generic and traditional specialties (clipfish, stockfish, and salted fish) (NSEC 2006; Xie and Myrland 2010). Icelandic and Faroese suppliers are nevertheless supplying many different cod preparations, more convenient for Spanish households. Norwegian interviewees agree in reporting that better fish quality is only achievable if the landed cod has been handled and stored under appropriate conditions. In other words, it appears to be necessary that fishermen's skills be enhanced in order to turn the convention of identity from abstract to personalized.

A regionalized approach to the Spanish salted cod market (see Espinosa, this volume, for further discussion on this topic) unveils a critical outcome: Norwegian cod is still preferred to other competitors' mainly in southern Mediterranean provinces like Valencia, Alicante, and Murcia, where consumption patterns are attached to domestic qualities like the taste of traditional dried and salted cod and the fishmongers' trustworthiness. Icelandic and Faroese cod deliveries perform much better in wealthier regions (Madrid, Catalonia, Basque Country, Navarre) where whiteness and freshness are much more influential (see Sect. 2). This geographical divide underpins the criticism about the inability of the Norwegian innovation system to support adaptation to the Spanish market, and to highlight a new set of qualities for salted cod. Problems are apparent in following market trends for convenience packaging and user-friendly preparations, that is, market qualities. But the FVC as a whole hardly emphasizes the potential domestic, civic, and ecological qualities. Fashionable values linked to wild nature, Norwegian origin, and fair revenue distribution via the quota allocation are not properly addressed, so Norwegian salted cod is increasingly pushed to the margins of the retail system and often overlooked as an anonymous staple.

It is likely that capitalization on different qualities requires agreement through the value chain and the surrounding institutional network about different conventions

Table 4.1 Seafood qualities in the Spanish market and challenges for Norwegian salted cod FVC. (Source: author's elaboration)

Orders of worth	Qualities in the Spanish seafood market	Challenges to Norwegian salted cod FVC
Market	Freshness Whiteness User-friendly packaging	Product diversification Market orientation
Industrial	Efficient value chain Nation-wide availability Frequent consumption	Vertical integration Investment in value-adding
Domestic	Fishmongers: service and advice TV and chefs: Mediterranean diet Supermarkets: in-house fish shops	Emphasis on geographical origin Focus on interpersonal retail outlets
Public	Attachment to brands	From bulk to branding strategy
Civic	Respect for rules and norms Seafood's healthiness No food scares involving fish	Scandinavian tradition of fairness De-salted and light-salted products Long-term agreements
Ecological	Aquaculture: more white fresh fish Market divide risk	Emphasis on ecological concern Leadership in aquaculture Focus on new market niches

related to resource management, quality standards, distribution systems, market targets, and a shared narrative that sends a clear and convincing message to consumers. But mistrust within the salted cod industry raises barriers to most forms of collective action (Lindkvist 2010). A new strategy 2010–2013 has been launched by the Norwegian authorities to encompass all players, but it is narrowly focused on salted and dried cod, whose limitations are increasingly apparent. Many Norwegian producers have shifted to lower-price countries like Portugal (Lindkvist, this volume), which actually imports 40 % of Norwegian *clipfish* (Xie and Myrland 2010; Xie, this volume) due to the achievement of a single quality standard enacted by the national government which fits both local preferences and Norwegian supplies.

4.5 Conclusions

The most important qualities that rule the Spanish seafood market, and their implications for Norwegian salted cod producers and for their supportive environment, are summarized in Table 4.1. Broadly speaking, Spanish consumers mostly welcome seafood endowed with commercial, domestic and civic attributes. Fish quality is mostly assessed by its freshness and whiteness, and customers demand information from their fishmongers about origin and recipes because fish intake is a healthy habit, deeply rooted in the Mediterranean diet, whose benefits are highlighted by food authorities and mass media. Therefore, seafood delivery throughout urban and rural Spain is the duty of an efficient value chain where new players, such as supermarkets and fish farms, seem to follow the overarching market trends and build in-house fish counters or supply the most appreciated fresh white species.

Some critical conventions that sustain the Norwegian salted cod FVC appear to be in fundamental disagreement with the Spanish context. Norwegian cod undergoes a standardized technological process that sources bulk salted cod splits (*butterflies*) to a generic market that is, in fact, increasingly differentiated. Since quota-setting and fish procurement are the key concerns for Norwegian FVC partners, their convention of identity is abstract or anonymous: quantity of landings is much more important than their quality. Thus, corporate or geographical branding is lacking, whereas Spain is very sensitive to trademarks and food origin. Engagement of the Norwegian authorities with marine and coastal affairs is praiseworthy in civic and ecological terms, but it is also misleading from a market perspective because it does not help producers to gain competitiveness in the more demanding foreign countries.

Nevertheless, some Norwegian conventions and practices might be capitalized to match the qualities demanded in the more profitable niches of the Spanish seafood market: geographical origin, social welfare, and environmental sustainability. First, Norway enjoys a very good reputation as a fishing nation (public qualities). Moreover, the salted cod industry is mostly located from the Møre and Romsdal region and northwards to Lofoten, so domestic qualities could be also invoked to turn the convention of identity from its current anonymity into a more personalized link to small local producers. Second, the quota system implies a restriction of the convention of participation which clearly points to a pretty fair distribution of income (civic qualities). Third, the commitment of a national innovation system with the sustainability of fisheries is still unknown to 'green' and affluent Spanish consumers, who are somewhat reluctant to adopt aquaculture. Fieldwork conducted in Spain by all researchers involved in this project has recorded the embeddedness of Norwegian salted cod outside the mainstream mass retail system, where qualities other than price and convenience may be easily transmitted to consumers. This small but not negligible position might be a starting point for a wider and more ambitious research and marketing strategy focused on these distinctive assets of Norwegian salted cod.

References

Boltanski L, Thévenot L (2006/1991) On justification: economies of worth. Princeton University Press, Princeton

Cruickshank J, Lysgård HK, Magnussen ML (2009) The logic of the construction of rural politics: political discourses on rurality in Norway. Geogr Ann Ser B Hum Geogr 91:73–89

DATACOMEX. Estadísticas del Comercio Exterior de España. http://datacomex.comercio.es

Eden S, Bear Ch, Walker G (2008) Understanding and (dis)trusting food assurance schemes: consumer confidence and the 'knowledge fix'. J Rural Stud 24:1–14

Gago JM (2007) El pequeño comercio en la posguerra castellana. De la cartilla de racionamiento a los supermercados. Junta de Castilla y León,Valladolid

Household Food Consumption Database. [Online]. http://www.mapa.es/es/alimentacion/pags/consumo/BD/consulta.asp

Ilbery B, Morris C, Buller H, Maye D, Kneafsey M (2005) Product, process and place. An examination of food marketing and labelling schemes in Europe and North America. Eur Urban Reg Stud 12:116–132

Kristoffersen B, Young S (2010) Geographies of security and statehood in Norway's 'Battle of the North'. Geoforum 41:577–584

Lindkvist KB (2001) Governance and territoriality in Norwegian fisheries. Nor Geogr Tidsskr 55:9–16

Lindkvist KB (2010) Mistrust and lack of market innovation. A case study of loss of competitiveness in a seafood industry. Eur Urban Reg Stud 17:1–13

Lindkvist KB, Sánchez JL (2008) Conventions and innovation: a comparison of two localized natural resource-based industries. Regional Studies 42:343–354

Lindkvist KB, Gallart L, Stabell MC (2008) The restructuring of the Spanish salted fish market. Can Geogr 52:105–120

MERCASA (2011). Alimentación en España 2010. Producción, industria, distribución y consumo. MERCASA, Madrid

Ministerio de Agricultura, Pesca y Alimentación (2007) Productos del Mar. Monográfico. Observatorio del Consumo y la Distribución Alimentaria. MAPA, Madrid

Ministerio de Agricultura, Pesca y Alimentación (2008) Hábitos de compra, conservación y consumo de los productos pesqueros en la población española 2007. MAPA-FROM, Madrid

Ministerio de Medio Ambiente y Medio Rural y Marino (2008) La cadena de valor del pescado fresco en España. MARM y Asociación de Cadenas Españolas de Supermercados (ACES), Madrid

Ministerio de Medio Ambiente y Medio Rural y Marino (2009). Consumo de Pesca. Panel de Hogares. MARM, Madrid

Morgan K, Marsden T, Murdoch J (2006) Worlds of food. Place, power, and provenance in the food chain. Oxford University Press, Oxford

Murdoch J, Marsden T, Banks J (2000) Quality, nature and embeddedness: some theoretical considerations in the context of the food sector. Econ Geogr 76:107–125

Norwegian Ministry of Fisheries and Coastal Affairs (NMFCA) (2010) Facts about fisheries and aquaculture 2009 NMFCA, Oslo

Norwegian Seafood Export Council (NSEC) (2006) Statistical overview of Norwegian seafood around the world. NSEC, Tromsø

Parrott N, Wilson N, Murdoch J (2002) Spatializing quality: regional protection and the alternative geography of food. Eur Urban Reg Stud 9:241–261

Ponte S, Gibbon P (2005) Quality standards, conventions and the governance of global value chains. Econ and Soc 34:1–31

Sánchez-Hernández JL, Aparicio-Amador J, Alonso-Santos JL (2010) The shift between worlds of production as an innovative process in the wine industry in Castile and Leon (Spain). Geoforum 41:469–478

Statistics Norway (2009) Fishery Statistics 2007. Statistics Norway. Oslo

Statistics Norway [Online] http://www.ssb.no/ur_okonomi_en/. Accessed 9 March 2011

Storper M, Salais R (1997) Worlds of production. The action frameworks of the economy. Harvard University Press, Cambridge

Xie J, Myrland Ø (2010) Modeling market structure of the Spanish salted fish market. Food Economics—Acta Agriculturae Scandinavica—Section C 7:119–127

Chapter 5
Regionalism in the Salted Codfish Market in Spain

Ana Espinosa Seguí and Inmaculada Martínez Alba

Abstract Spain has become a larger importer of salted codfish due to the drop of its own national production throughout the last decades. Since then, international players in the Spanish salted codfish market have been introducing their salted codfish in order to buy up the Spanish market, with good marketing strategies aimed at changing the traditional idea of salted fish quality. On the demand side, Spain is not a homogeneous market. The different behaviour of the main consumer regions modulates important aspects of domestic consumption, creating a mosaic of smaller markets. This mosaic is enhanced by the presence of regional and local producers, deeply linked with their consumers.

Keywords Salted codfish · Mature product · Market changes · Concept of quality

5.1 Introduction

Gastronomy is one of the most important and representative elements of the culture of any society, since it is able to show how people produce, distribute, prepare and consume its sustenance, following internal rules and conventions agreed for the majority of agents of a particular society (Banks 2000; Straete 2004, 2008; Ponte and Gibbon 2005; Lindkvist and Sánchez 2007; Sánchez et al. 2010).

Very commonly, local and regional gastronomy have been linked to those products produced or raised in a close and small market, with a local or even, regional scale (Straete 2008). However, Spanish gastronomic culture has been enriched for centuries with other products that were not originally native to the country, but deeply rooted in Spanish gastronomy and culture, such as salted codfish (Gallart et al. 2003, 2005a).

From the fifteenth century, Basque fishermen started the trade in Spain of codfish caught in the North Atlantic Area (Gallart et al. 2003; Richter-Hanssen Chap. 3).

A. Espinosa Seguí (✉)
University of Alicante, Alicante, Spain
e-mail: Ana.Espinosa@ua.es

I. Martínez Alba
University of Seville, Seville, Spain

K. B. Lindkvist, T. Trondsen (eds.), *Nordic-Iberian Cod Value Chains*,
MARE Publication Series 8, DOI 10.1007/978-3-319-16405-2_5

Cleaning, preparation and preservation with salt on ship-board of the fresh codfish resulted in a strong flavoured and golden coloured product. This salted codfish became very popular all over Spain but mostly, among inland consumers, with problems of access and low purchasing power for buying fresh fish caught in the harbours of the country (Gallart et al. 2005a; Richter-Hanssen Chap. 3).

Thus, salted fish became a staple prepared in many different ways around the country, with important peak consumption during Christmas and mostly, Lent, when the Catholic precepts did not allow consuming meat among parishioners (Kurlansky 1998; Gallart et al. 2005a; Sánchez 2011).

Throughout centuries, codfish has gained popularity and become more established in the culture among Spanish consumers, being used in many dishes and even, in proverbs, as an element of Spanish traditions. So much so that salted codfish is considered for many consumers and producers as something different from meat and all other fishes, and it lends it self to a really wide collection of regional recipes.

Although Spain has been seen as a particularly important market for countries engaged in salted codfish production, there exists in fact a mosaic of smaller markets with regional rooted conventions (Cornish 1997; Gallart et al. 2005a). In some cases the salted fish is exported to the other regions and widely used by their populations. In some cases, however, many people are so traditional and locally oriented that the products are unknown outside the borders of the city, province or region. Moreover, these conventions do not correspond with the administrative regionalisation of the country, and sometimes it is possible to observe different regions with more similarities in the desired taste and way of producing salted codfish than other areas of the same region.

The importance of salted codfish in both the offer and demand sides of Spanish markets has motivated a research for understanding the behaviour of different submarkets, their hinterlands, preferences of codfish products and desired countries of origin. For this reason, the objective of this chapter is to show the main features of salted fish regionalised markets.

5.2 Methodology

Since the main objective of this research was to find information that corroborated the internal division of the Spanish market among regions and through the last decade, it was necessary to compare quantitative data with qualitative information gathered through in-depth interviews of the main agents of the salted codfish market.

Starting with quantitative information, the main sources for comparing the recent evolution of the salted fish Spanish markets were regional statistics, and *Datacomex*, which is the Spanish foreign trade statistics service. This information from the offer side, enriched with quantitative data gathered through polls, was sent to the salted codfish producers sited in Spain. The quantitative information on the demand side was based on statistical data from the *panel de consumo* (home consumption database), the national consumption panel, published by MARM, the Ministry of Environment, Rural and Marine Spaces.

Table 5.1 Qualitative information of the salted codfish market. (Source: Fieldwork made in the above proposed regionalisation of Spain by the authors)

Regions	Companies operating in 2009	Interviewed companies
Galicia	9	6
Basque Country	26	13
Catalonia	8	6
Region of Madrid	3	3
Levante	11	8
Andalusia	5	5
Total	*62*	*41*

Qualitative information completed the data gathered from national and regional statistics and was helpful for understanding the regional behaviour of producers, retailers and consumers (Table 5.1).

The in-depth interviews were split into two main groups: the supply and the demand sides. Whereas the supply side was covered only with interviews of national producers or importers, excluding in-country of origin producers, the demand side was made up of retailers, consumers and restaurants. The group of retailers has much information about the markets and conventions established at any spatial scale, being able to influence much more the final consumers, both at home and in the foodservice sector. At the same time, and being always a very popular and traditional product, salted codfish consumption has evolved differently at home or in bars or restaurants. It was therefore important to compare trends and conventions used in the most representative consumption places.

Building on the preliminary hypothesis that the Spanish market is divided into many different submarkets, with regional conventions established by the main agents involved in the market, the research was focused on six different regions, which do not coincide with the administrative borders of Spanish regions (Fig. 5.1).

In this sense, we studied together Galicia and Cantabria, the Basque Country with La Rioja and Navarre, Aragon together with Catalonia. The Region of Madrid and Andalusia were studied alone because of the size and independence of the markets and finally, the Region of Valencia and Murcia were studied together as the Levante region[1].

5.3 Evolution of the Salted Codfish Market in Spain

A first approach to the demand of cod from the perspective of imports reveals the diversity of preferences that exist at regional level (Xie and Myrland 2010). In this sense, in the Levante Region and Catalonia, the category "salted without drying or

[1] As *Levante* we consider the southeast part of the Iberian Peninsula, including regions of Valencia and Murcia.

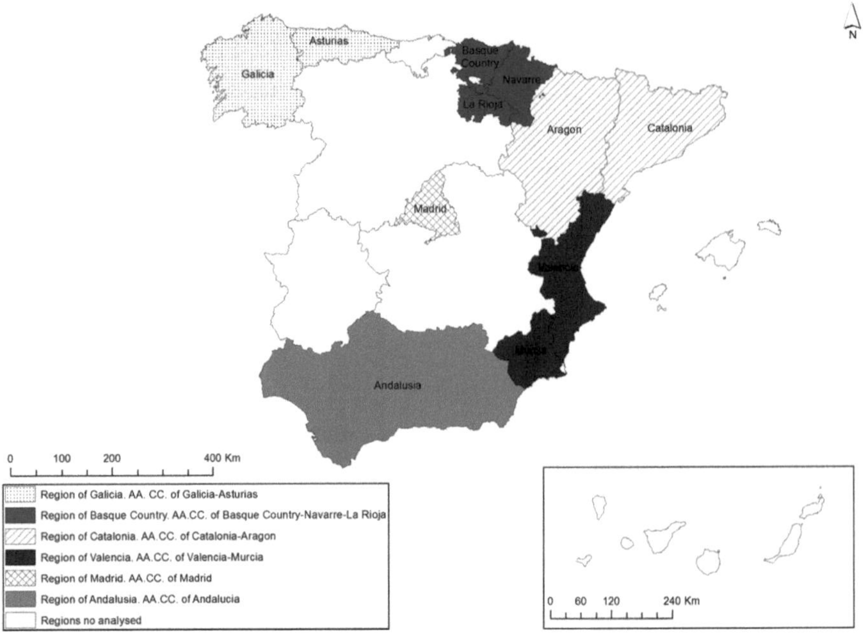

Fig. 5.1 Regions analysed in this research. (AACC: Autonomous Communities. Source: Authors)

smoke, also in brine" is the most important in terms of volume, since this is the basic salted codfish product produced in Spain (Table 5.2). This category is also called "*Bacalao verde*" and makes reference to half cured salted fish, which is salted again, sometimes also dried and cut into pieces for final consumers.

On the other side, in Galicia, it is important to highlight that frozen codfish has become insignificant, whereas the *bacalao verde* category, or salted without drying or smoke, also in brine, is the leader of the salted fish products (Table 5.2).

In Madrid, fresh codfish has duplicated its volume throughout the last 10 years, but as it happens in Galicia, *bacalao verde* is the indisputable leader of the salted codfish categories; even volume is half of the fresh fish imported in Madrid (Table 5.2). Also in the Basque Country, the fresh codfish is still the front-runner in imports. Finally, in Andalusia, the principal in relation to the imports came from dried salted cod.

Of course, apart from the official statistics offered by Datacomex, it is absolutely indispensable to take into account that the internal trade market of salted codfish in any variation changes the import/export Spanish map. In this sense, Basque Country, Madrid and Galicia act as the main suppliers for regions as Murcia, Valencia, and the inland area of the country.

As regards the origin of imports, it is necessary to point out general features. Thus, in the period 2000–2008, the import of Norwegian codfish has declined and *Bacalao nacional*[2] (Figs. 5.2 and 5.3), became the second largest source of cod in the State (see Chap. 7).

[2] Captured by the Spanish codfleet, salted on board and cured for a period of at least 4 months.

Table 5.2 Direct importation of salted codfish in Spanish regions (in tons). (Source: datacomex.comercio.es/)

Region	Year	Fresh	Frozen[b]	Chilled and fresh filets[b]	Frozen filets[b]	Dried salted	Salted without drying or smoke, also in brine
Galicia	*2000*	255	66	–	1	27	161
	2009[a]	404	18	24	–	2	359
Basque Country	*2000*	982	467	882	66	250	4199
	2009[a]	679	81	–	–	327	4677
Catalonia	*2000*	623	–	322	1819	–	4.427
	2009[a]	1043	1404	–	–	22	3.141
Madrid	*2000*	95	–	128	69	–	254
	2009[a]	225	0	–	–	–	120
Levante	*2000*	75	43	126	–	367	1102
	2009[a]	27	0	–	–	44	2530
Andalusia	*2000*	12	–	–	–	–	90
	2009[a]	41	45	–	–	132	–

[a] Provisional data
[b] Data from *Gadus morhua*, *Gadus macrocephallus* and *Gadus ogac*

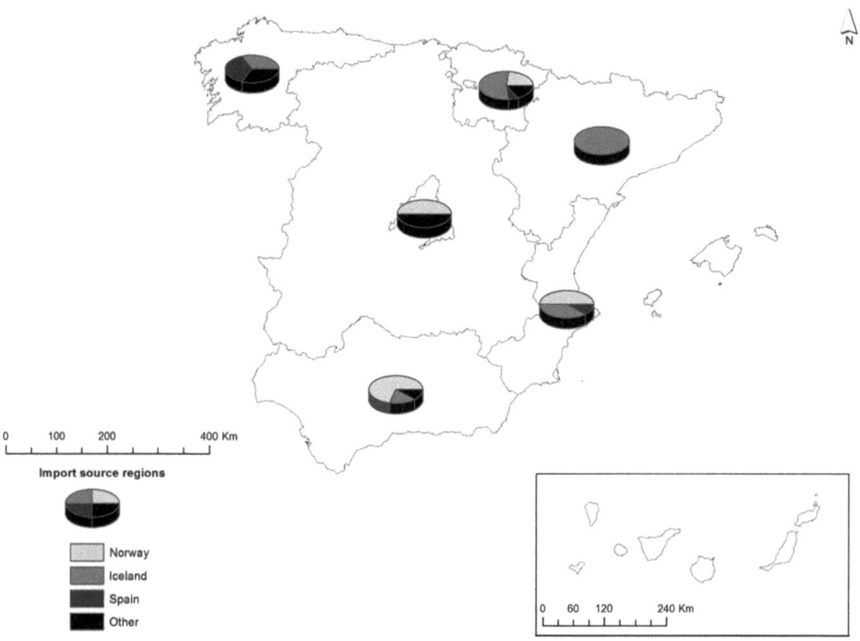

Fig. 5.2 Countries of origin of imports (2000). (Source: www.datacomex.comercio.es/ Accessed March 2011)

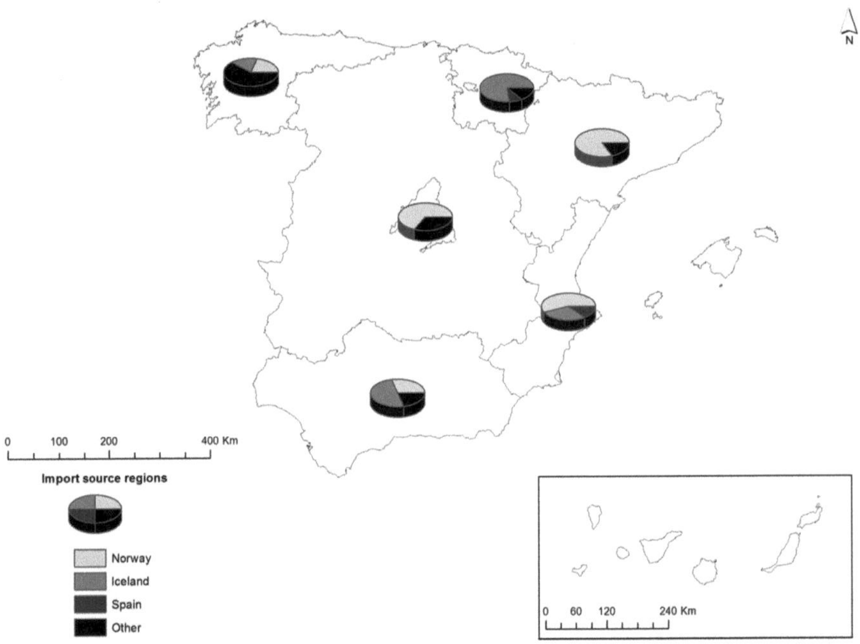

Fig. 5.3 Countries of origin of imports (2008)

Table 5.3 Home salted fish consumption in Spain.(Source: MARM 2009) (Since consumption of salted fish is very low in comparison with fresh or frozen fish in Spain, there are no disintegrated data for just salted codfish. For this reason, consumption data published by the Ministry of Agriculture, Food and Environment gathers any kind of salted fish consumption under the same category)

Years	Volume (t)	Value (1000 €)	Average Price (€ /kg)	Consumption Kg per capita
2005	22,898	195,934	8.56	0.54
2006	21,791	209,266	9.6	0.51
2007	19,414	198,058	10	0.44
2008	18,206	204,940	11	0.41

However, throughout this decade, the geographical map of codfish supply in Spain has dramatically changed. First of all, the decline in catches of *bacalao nacional* explains the increase in imports. In the case of Norwegian codfish, the decline of importation was related to the rise of imports from other countries, mainly Iceland and the Faroe Islands (Lindkvist et al. 2008). Regarding the case of Iceland, this country implemented an aggressive marketing campaign for its products, which included high quality standards for the product and customer service.

On the demand side, salted fish consumption in Spain, including any kind of salted fish, such as anchovies, sardines, tuna, codfish or other pelagic salted fishes and every grade of cure, has in general declined over the past 4 years, although the value of these products has increased (Table 5.3). Even so, the European Commission (2009) claimed that *"demand for dried, salted cod remains strong in the Iberian peninsula"* based on the observation that 25 % of the Spanish population consumes dried salted cod every 2 or 3 weeks (MMRM 2009).

The trends in *salted codfish* consumption patterns can be understood in a context of the larger economic, social and cultural changes taking place in Spain (Grafe 2004; Espinosa and Martínez 2011). One of the producers expressed it like this: "*sales of salted fish have dramatically fallen down along the last two decades* [...] *People don't want to prepare salted fish. They don't have the time nor the expertise, so who would like to do something when you can buy directly for almost the same price?*".

At the same time, a new market segment of cheaper products has evolved in parallel to the traditional salted fish market, not just in the Basque Country but in the whole country, focused on quick and convenient desalted or lightly salted frozen products (Table 5.4). Producers have adapted their supply in order to fulfil the new expectations of the market. A big manufacturer of Navarre expressed it like this: "*Frozen codfish has gained a large market share not just against traditional salted codfish, but also against other frozen fishes. Traditional salted fish is decreasing its presence in the markets (...) there is not generation change in the sector*".

5.4 Regional Markets

5.4.1 Regionalism in Spain

Regional conventions in space and time in Spain have facilitated a much regionalised market for salted fish. Spain is thus not only one common market for salted codfish but consists in the addition of many market segments of different tastes and preferences for the same product (Gallart et al. 2003). A more detailed study of six different consumer regions will in the following subsections be analysed based on our fieldwork and research.

5.4.2 The Basque Country

Salted codfish has been an essential in development of the Basque gastronomy. Pasajes, in the province of Guipuzcoa has, apart from having been the leading harbour for the Spanish cod fishing fleet for a long period, also became the main entrance of imported salted codfish in the country (Espinosa and Martínez 2011). Thus, Basque people developed an incredible tie to salted codfish, extolling it as one of the most important foods of the Basque cuisine (Kurlansky 1998; García 2007).

Nowadays, although the cod fleet has moved to Galicia (Lindkvist and Lois 2000), Basque importers, producers and distributors have developed Spain's densest cluster of codfish, supplying the entire country and as a reference in the sector in the country

Table 5.4 Regionalisation of salted codfish. (Source: authors). Based on Murdoch and Miele 1999; Ponte 2009

	Competitive environment	Tradition versus innovation	Market area	The best codfish is...
Basque Country	Densest cluster of producers and importers	Basque chefs are known trendsetters who combine tradition with new recipes based on cod	The whole country and Portugal as well as other international markets	Between 2–3 kg, long-line fished, able to laminate, long cured and not injected
Region of Madrid	Scarce number of producers but a very big market	Madrid is the most representative Spanish gastronomic melting pot	The big, regional market of Madrid and mostly the south of the country	Original from Faroe Islands, Iceland or Norway, long cure time and long-line fished
Galicia	Dense cluster of fishermen, producers, importer and distributors	Galicia retains traditional features of consumption and distribution	Mainly focused on the regional market and Portugal	Very dried, no white coloured, not injected, long cured
Catalonia	Dense net of big importers, distributors and cooperatives of retailers ("*bacalladers*")	Introduction of new conventions in Catalonia and imposing these new trends to the rest of the country	Catalan market bought up by regional producers and in search of new markets in the country	Big, thick loins, injected, not long cured, very white and well soaked
Andalusia	Concentration of producers in very few but big companies for (inter)national markets	Traditional and low purchasing power market	The big but regional market of Andalusia	Very dried and salted, long cured and small sized
Levante	Regional and traditional companies, of "*salazoneros*"	Traditional market not adapted to new conventions	Regional market restricted by regional conventions	*Bacalao nacional*, goldened colour, very cured for raw consumption (*tipo inglés*)

Bacalladers is an old trade in Catalonia which consists in distributing and selling ready to use, soaked codfish without bones and skin

In the Spanish Levante, salted fish production, commercialisation and consumption has been so historically important that manufacturers of the region do not consider themselves as codfish producers, but *salazoneros*. *Salazón* includes any type of salting fish from codfish to sardines, octopus, tuna or mackerel

and the neighbour markets. For this reason, the Basque Country is considered one of the most important submarkets of Spain for salted codfish, both on the supply and demand sides.

On the supply side were 26 producers and importers working in 2009. Basque producers have a lot of information about the markets they control, not just within their boundaries but also about the national demand. Basque "demand" represents consumers willing to pay more for high quality and traditional products. For this reason, the bulk of Basque manufacturers have introduced new and more convenient products as lightsalted and frozen desalted codfish products.

Basque producers do not like this frozen desalted product, especially those who have been working in the sector for a long, long time and are used to hard taste and well salted products. "*Light-salted fish is the cheap version of salted fish, but it is not the same… I always explain the differences to my clients, but they do not mind. Those who is looking for a low price, really doesn't mind about the quality and the organoleptic features of codfish*".

On the demand side, the influence of salted codfish in the foodservice sector has been high. Trendsetters like well-known chefs and many bars and restaurants have included many different dishes of salted codfish in their menus. Cider bars are one of the best examples of the embedding of salted codfish in the Basque cuisine.

5.4.3 The Region of Madrid

"*Madrid is 350 km from Valencia, 393 from Santander, 395 from Bilbao, 420 from Alicante, something more than 600 from A Coruña and around 660 from Cádiz* (Bernardos 2001)". Madrid is the capital of the country, sited in the centre of the Iberian Peninsula, with no direct connection with the sea. However, Madrid is considered the biggest fish market of Spain and Europe and the second largest of the world, after Tokyo (Espinosa and Martínez 2011). Due to its excellent net of roads, chefs and wholesalers of the whole country travel to Madrid to find the best fresh fish products in Spain, including seafood, even those located in fishing harbours of the stature of Vigo, San Sebastian or Barcelona.

Also, due to its strategic location, Madrid has become a meeting point for producers and distributors, which helps the region to become a big market with a regional offer and demand. The supply side is made up of three big companies, which supplies salted codfish to the whole region and provides the southern part of the country with codfish adapted to its conventions.

These three companies are also direct importers of salted codfish from countries with a high volume of production, such as the Faroe Islands, Norway, Iceland and Russia. The first three countries have bought up the market in Madrid for medium and high quality, whereas Russia as a new player, has fulfilled the needs of the cheapest part of the market.

Although the number of suppliers in Madrid is low, the market is one of the more compact of the whole country (Espinosa and Martínez 2011). Due to its high density of population, Madrid has a large volume of consumers, taking into account consumption at home and in the foodservice sector. In fact, the group of foodservice companies of Madrid is made up of a melting pot of many regional gastronomic influences from the whole Iberian Peninsula. For this reason, Madrid shares many recipes with other regions that have a long tradition of salted codfish consumption, being a good example of the variety and richness of salted codfish based Spanish cuisine. At the same time, Madrid has been a good test market for new products as light-salted codfish or desalted and frozen codfish, since it has a large number of urban consumers, willing to purchase more convenient and easy to prepare products, rather than traditional salted codfish.

5.4.4 Galicia

Together with the Basque Country, Galicia has had the closest contact with capture and production of salted codfish. Indeed, its regional economy has the largest participation of fishing activities in its whole country (Pérez 1985; Lindkvist and Lois 2000) and both culture and society are influenced by its closeness to the sea.

The proximity to Portugal has also had an indisputable influence on the gastronomic conventions of Galician society, resulting in salted codfish consumption in Galicia sharing many aspects in common with the neighbour country (Espinosa and Martínez 2011). This relationship between them is so extreme that the closer the distance; the higher the similarities among places in both countries. As one producer commented: "*We have different clients according to the closeness to Portugal. Whereas people from Pontevedra want very dried and salted fish, consumers in A Coruña do prefer less than 3/4 cure*".

This proximity to Portugal has influenced the regional consumer conventions of Galicia, which is one of the most traditional regions of the country. Salted codfish in Galicia must be very hard salted and dried, traditionally manufactured and sold as a whole fish. For this reason, the main manufacturers of Galicia have not been interested in modern products, at least for their regional market. One traditional manufacturer declared that "*I am just specialised on traditional salted codfish, not anything else. . . . It is possible that in the future other products as chilled or frozen desalted will gain more market, but Galicia is a very traditional market. . . . Just an example . . . I have to cure and dry more fish in the factory, since it comes in as salted but not dried. We like the fish as the Portuguese taste, very dry and cured*".

This traditionalism is also seen in the distribution channel and in the way consumers prepare the codfish at home or in the foodservice channel. In this sense, Galician retailers usually sell whole split codfish, with no more handling than the salting and packaging processes, directly to local producers. This means that consumers are highly skilled in cutting and desalting fish as they have been doing for many years. At the same time, Galicia does not have the problems that other regions in Spain are suffering, like high differences in the standards of quality, price and mainly, display between traditional and modern distribution channels.

5.4.5 Catalonia

Through the last decades, manufacturers from Catalonia have increased their participation in the salted codfish value chain. By controlling importations, production settled down in Catalonia and enlarged their market hinterland towards other regions of Spain (Espinosa and Martínez 2011). Traditionally, Catalonia has been a large market for salted codfish landed in Galicia or Basque Country with important regional industrial conventions focused on the preparation of codfish. In this sense, Catalonia has been famous within the codfish sector for its salted codfish processors association, called "*bacalladers*" in Catalan, who are retailers or distributors skilled in desalting codfish or taking the bones and skin before selling it to the final consumers. *Bacalladers* have always been tightly linked to the market, fulfilling housewives and chefs' criteria for good codfish and helping them with the arduous task of salted codfish preparation.

In recent decades, Catalonia has evolved towards a more complex salted codfish market, not just focused on the final clients, who are the absolute leaders, but also aimed at the supply side of the market. The participation of *bacalladers* and small retailers is so important in consolidation of the market that this region has the only cooperative of salted codfish producers, consisting of small retailers, most of them sited on market stalls, corner shops or even, street markets.

Also, the long tradition of selling codfish soaked and without skin or bones has changed the regional consumption conventions, being based on white colour, thickness or big loins (Gallart et al. 2005b). For this reason, manufacturers and importers of salted codfish have looked for Icelandic producers, able to supply the required needs of retailers and consumers in Catalonia. Our fieldwork in Catalonia led very quickly to the totality of Catalan companies trusting Icelandic fish for their own market, as well as introducing Catalan conventions to the other Spanish regions: "*My family always bought the salted fish from Basque producers and fishermen, but now, Basque fishermen do not have so much influence on the market.... Now, the whole market moves towards imported fish, mostly from Iceland. Catalonia has understood very well the new trends, and the rest of Spain will follow us.*"

Whereas in other regions participation of companies from different countries of origin of salted codfish is more equal, always taking into account regional conventions and preferences, Catalonia has been bought up by Icelandic codfish and manufacturers of both countries have very good connections between them. As one manufacturer said: "*Our company imports salted codfish directly from different countries of the North of Europe, basically Iceland, since its industry has shown a very good adaptation to the Catalan taste and demand. Iceland is the Spanish main salted fish supplier, selecting and manufacturing in its factories everything which is necessary for a display of the final product.*"

Since Catalonia is considered an important market in terms of purchasing power and volume of consumption, salted codfish companies have had a very good performance in the region and this success in Catalonia has boosted the enlargement of Catalan companies' market hinterland, in the southern part of the country (Espinosa and Martínez 2011). It would imply diffusion of Catalan conventions into other regions and the breaking down of regional conventions, which is difficult to see according to the regionalisation of salted codfish consumption. This is the one of the most important objectives of Catalan companies, as many producers admitted. Quoting one interviewee: "*in the rest of Spain, there is still a traditional conception of the codfish consumption, but I am sure that along the next years, every market of Spain will be driven by Catalan salted cod conventions* [...] *The market is already established with these new orientations: white, thick codfish.... Catalonia has totally embedded these new trends and we are all happy with them*".

5.4.6 Andalusia

The concentration of production linked to a strong commitment to investment oriented manufacturing high quality products had led to the disappearance of manufacturing companies, having the effect of halving the number of companies in just 5 years. So, today the cod salting industry in Andalusia is made up of five companies (Espinosa and Martínez 2011).

These companies mainly supply cod from China, Iceland and the Faroe Islands. Several businessmen point out "*... every year I receive product and price catalogues from Iceland; moreover we are in almost permanent contact via email*". The sales area is regional basically, but it is undergoing a process of internationalisation by means of the introduction of these products in the Japanese market.

The preference for a constant supply, facilities for bargain prices through a single intermediary in the country of origin and the color "white" of cod, are the aspects that have caused import of cod to tilt from Norway to other countries of origin. An often heard comment was "*The best cod is that from Norway, but consumers do not like yellow cod*".

Most imports of codfish go through a half healing process; however there are also imports of frozen cod to prevent fish from being injected (Lauritzen 2004). According to one company: "*We prefer the average half salted; Spanish consumers like the fact of desalting at home*".

The salting codfish industry of the region has an artisanal character. However, in recent years it has shown a strong commitment to research and development in processing. Thus, the regional industry has directed its efforts in recent years towards the market introduction of new products (Espinosa and Martínez 2011).

The supply of salt cod from Andalusian companies is very diverse, despite the best-selling product being traditional dry salted cod. The Andalusian salted cod products have experienced in the past 5 years a high level of dynamism. In this area, highlights were average price increases of these products, and higher billing to companies (Consejería de Agricultura y Pesca 2010).

The producers used their own brands to market their products, but also sold to wholesalers for distribution in the market under private labels. This led the Andalusians to cater to consumer preferences for food products by a quality side and another low cost.

The consumption of salted codfish at home has slightly diminished in Andalusia in the last 5 years (Consejería de Agricultura y Pesca, 2010). Nevertheless, its value in the market has increased, which is an aim for the revaluation of the product because of the enterprise strategies that have managed to introduce salty codfish in the product niche of "quality" or "discharge range".

Salt codfish has become an essential product in the spring, due to the importance of cultural or religous factors in the regional cuisine.

5.4.7 The Region of Levante

In comparison with other regions such as Catalonia, Galicia or the Basque Country, the Spanish *Levante* has the longest tradition in both consumption and production of any sort of salted fish. On the opposite side, the structure of the companies sited in the region is more traditional, family based and small sizes, except one of them, which is the biggest one of Spain (Espinosa and Martínez 2011).

In *Levante*, codfish is part of the salted fish tradition and gastronomic richness (Gallart et al. 2005a) but it has, for many reasons, never held a leading role: a very specific way of consumption, preparation in small portions or the existence of other traditional salted fishes such as tuna, mackerel, octopus or sardine.

Nevertheless, the standard consumption of salted codfish is also as common as in the rest of the country, although few regional recipes and traditions are linked to this way of consumption.

Although salted codfish is deeply rooted in the gastronomy of the region, being an indisoluble component of the traditions and culture of both regions, Valencia and Murcia, it is perceived more as simply another kind of salted fish (Gallart et al. 2005a). Used for appetisers, snacks, baguettes or salads and consumed raw, salted codfish has a completely different use and perception in this region. For this reason, one producer said "*we are not bacaladeros (salted codfish producers), but salazoneros (salted fish producers), using many kinds of salted fish, mainly tuna*".

As a direct consequence of these close conventions, manufacturers of salted codfish in the region have a very small hinterland, always within the borders of the region. However, the North of the Valencia region is more influenced by the Catalan conventions, due to the great distance from the main places of production, sited in Murcia and Alicante (Espinosa and Martínez 2011).

Raw consumption of salted codfish is due to a special treatment in which salted codfish is washed and again cured. That provokes a very goldened colour, a strong flavour and less salt than the pre-desalted salted codfish. As this technique was introduced in the times when England controlled the codfish market, this particular codfish was coined in the region as "*bacalao tipo inglés*" (English style codfish).

The market hinterland is very scarce due to the exclusivity of regional conventions, but in the last years, also production in the region has suffered from a shortage of specific raw material. The progressive abandonment of *Bacalao Nacional,* produced by the Spanish cod fleet, has made the production more difficult of *bacalao tipo ingles* in the region. This means that regional conventions have not changed but raw materials have, since they come mainly from other countries such as Iceland and

Norway and not from the national production. For this reason, many producers have tried to maintain regional conventions by treating the new codfish with traditional techniques that colour the fish. As a retailer from the Market of Alicante, a *salazonero* and distributor with long experience in the salted fish market said "*Yellow and golden colours are so important here that even manufacturers add sometimes saffron, food colouring or paprika to get a goldened colour in the codfish that they are selling*".

5.4.8 Comparing the Spanish Cod Markets Jigsaw

After the previous regional analysis, it is possible to conclude that Spain can not be considered as a homogeneous market for salted codfish. The long tradition of this product, its versatility in the local and regional gastronomy and the geographical and historical background of any of the regions analysed, have created a map of different tastes, preferences and recipes in the context of Spain.

According to Table 5.4, regions such as Basque Country, Galicia and Catalonia concentrate the bulk of salted codfish importers and producers of the whole country, being considered trendsetters in the establishment of both production and consumption conventions.

The regionalization of salted codfish is mostly highlighted in the column "*the best codfish is...*", since it is possible to observe how different preferences and conventions in the Spanish geography are.

5.5 Conclusions

After carrying out this research in the six salted codfish consumer regions, there is no point in discussing the embeddeness of salted codfish in the Spanish gastronomical culture. It is clear that there exists a regionalisation of production and consumption of salted codfish, and this fragmentation of the market is not absolutely linked to administrative boundaries but to former markets and similar cultural background in the territory.

Despite this cultural embedding of salted codfish in the country, during the last two decades the traditional patterns of consumption have evolved towards different niches. On the offer side, "*bacalao nacional*" is slowly disappearing from Spanish cuisines, due to abandonment of the Spanish cod fleet, which means the disappearing of a traditional and deeply rooted fishing activity. This fact has led to strengthening in the industry of Scandinavian countries such as Iceland, Denmark and Norway and the entrance of new players, such as Russia, China and USA (the state of Alaska). At the same time, the need for making cheaper priced salted codfish has invited the entrance of new fish products, similar in taste and aspect to codfish.

The National Association of Cod and Salted Cod Producers (ANFABASA)[3], which is associated with the bulk of salted codfish producers, was created to foster this sector's competitiveness by defending their interests and by promoting the consumption of salted cod and fish.

On the demand side, traditional products are suffering problems in finding rooms in many new Spanish homes, where inhabitants look for convenient and ready to eat products. That is the main reason for the consolidation of new products, such as chilled desalted, frozen desalted or light-salted codfish, more similar in their preparation than other fishes with a significantly lower price. Senior consumers or sybarites still delight in salted codfish consumption, as a unique, traditional dish. As far as these consumers will continue to demand traditional products and will encourage younger consumers to taste traditional and exclusive Spanish flavours, both markets of salted codfish will survive in parallel.

Acknowledgment Thanks to the research project "The Spanish salted fish market and opportunities for the Norwegians" and its leaders Knut Bjørn Lindkvist and Torbjørn Trondsen. Both researchers had the opportunity to research into the very interesting topic of food in Spanish gastronomy. Thanks also to the whole project team, especially José Luis Sánchez Hernández and Heidi Bjonnes Larsen who supported, encouraged and helped us in all our research.

References

Banks J (2000) Quality, nature, and embeddedness: some theoretical considerations in the context of the food sector. Econ Geogr 78:107–125

Bernardos Sanz, JU (2001) El abastecimiento y consumo de pescado en Madrid en el Antiguo Régimen. Vii Congreso de la Asociación de Historia Económica

Consejería de Agricultura y Pesca (2010) Análisis de la producción pesquera regional, año 2009, Resource document. Junta de Andalucía. Accessed 25 March 2009

Cornish S (1997) Product innovation and the spatial dynamics of market intelligence: does proximity to markets matter? Econ Geogr 73(2):143–165

Espinosa A, Martínez I (2011) Profiles of the Spanish salted fish markets. Working paper. Department of Geography, University of Bergen

European Commission (2009) Study on the supply and marketing of fishery and aquaculture products in the European Union. Directorate-Generale for Maritime Affairs and Fisheries

Gallart-Jornet L, Rodríguez-Barona S, Barat JM, Fito P (2003) Elaboración y comercialización del bacalao salado. Alimentaria 348:87–94

Gallart-Jornet L, Escriche Roberto I, Fito Maupoey P (2005a) La salazón del pescado, una tradición en la dieta mediteránea. Universidad Politécnica de Valencia, Valencia

Gallart-Jornet L, Heide M, Ostli J (2005b). Market situation of Norwegian bacalao on the Mediterranean Spanish coast. Okon Fisk 15:51–55

García Orellán R (2006) Rumbo al Gran Banco: una etnohistoria de la pesca industrial del bacalao en los bancos de Terranova. Rev Int Estud Vascos 51(2):577–592

García Santos R (2007) El bacalao en la cocina vasca. Iano books, San Sebastián

[3] ANFABASA website: http://www.anfabasa.com/. ANFABASA is a member of the *Asociación Nacional de Fabricantes de Conservas de Pescados y Mariscos (ANFACO)*, (http://www.anfaco.es/webs/webAnfaco/portales/anfaco/).

Grafe R (2004) Popish habits vs nutritional need: fasting and fish consumption in Iberia in the Early Modern Period. Discussion Papers in Economic and Social History, 55, May 2004. University of Oxford
Home consumption database (2009) Ministry for the Environment, Rural and Marine Areas (MMRM). http://www.mapa.es/es/alimentacion/pags/consumo/BD/consulta04.asp. Accessed 25 March 2009
Kurlansky M (1998) Cod: a biography of the fish that changed the world. Penguin Books, Westminster
Lauritzen K (2004) Quality of salted cod (Gadus morhua L.) as influenced by raw material and salt composition. Dr. Scient Thesis. Norwegian College of Fishery Science
Lindkvist K, Lois-González R (2000) Aspectos regionales de las pesquerías de Galicia y Noruega. Eria 51:69–77
Lindkvist K, Sánchez JL (2007) Conventions and innovations: a comparison of two localised natural resource-based industries. Reg Stud 42:343–354
Lindkvist K, Gallart L, Stabell MC (2008) The restructuring of the Spanish salted fish market. Can Geogr 52:106–120
Murdoch J, Miele M (1999) 'Back to nature': changing 'worlds of production' in the food sector. Sociol Rural 39:465–483
Pérez Fariña ML (1985) La industria de bacalao en Galicia: la importancia de la Ría de Arosa. Paralelo 37(89):425–444
Ponte S (2009) Governing through quality: conventions and supply relations in the value chain for South African wine. Eur Soc Rural Sociol 49(3):236–257
Ponte S, Gibbon P (2005) Quality standards, conventions and the governance of global value chains. Econ Soc 34(1):1–31
Sánchez JL (2011) 'The food value chain as a locus for (dis)agreement: conventions and qualities in the Spanish wine and Norwegian salted-cod industries'. Geogr Ann: Series B, Hum Geogr 93 (2):105–119
Sánchez JL, Aparicio-Amador J, Alonso-Santos JL (2010) The shift between worlds of production as an innovative process in the wine industry in Castile and Leon (Spain). Geoforum 41:469–478
Straete EP (2004) Innovation and changing worlds of production. Eur Urban Reg Stud 11(3):227–241
Straete EP (2008). Modes of qualities in development of speciality food. Br Food J 110(1):62–75
Xie J, Myrland O (2010) Modeling market structure of the Spanish salted fish market. Food Econ-Acta Agric Scand, Section C 7(2 & 4):119–127

Web Pages

Ministry of Environment, Rural and Marine Spaces (MARM) (2009) www.marm.es. Accessed 25 Feb 2009
Ministry of Economy and Competitiveness (2009) www.datacomex.es. Accessed 25 Feb 2009

Chapter 6
How do Green Orders of Worth Affect the Spanish Salted Fish Market?

Heidi Bjønnes Larsen

Abstract This chapter departs from the 'quality turn' concept, which is an academic term for consumer shift from a price-biased focus to an emphasis on non-economic aspects of food such as health, food safety, green values and animal welfare. Inspired by the 'quality turn', this chapter asks if Spanish consumers in the near future will require a more natural color for salted fish and resist the addition of artificial additives in the fish. Because color is essential to consumer definitions of quality and because color is influenced by the production process and the treatment of fish throughout the value chain, this chapter considers the color of salted fish to be an expression of quality. Based on secondary literature, telephone interviews with Spanish producers and interviews with owners of *sidrerias* (cider houses) in the Basque region, the chapter investigates possible changes in conventions toward a more natural white fish. The chapter investigates if there will be a change in the current quality convention in favor of green and civic conventions. The focus of the chapter is on supply, demand, and regulations in the value chain for Spanish salted cod.

Keywords Value chain · Quality understanding · Conflicting conventions · 'Quality turn' · Spanish market · Salted cod

6.1 Introducing Green Orders of Worth or the 'Quality Turn'

In the Western world, food scares and an increased focus on environmental issues are slowly changing consumer values about food choices. Scholars have referred to the turn away from price toward an increasing focus on noneconomic aspects of food such as diet and health, food safety, green values and animal welfare, as a 'quality turn' (Murdoch and Miele 1999; Murdoch et al. 2000). Scholarly examinations of the quality turn have focused on alternative food chains, but the quality turn also has consequences for conventional food value chains. Allaire (2010) sees the quality turn as an increased focus on quality conventions, pushed forward by market enlargement and a trend toward the individualization of consumers. The increased scholarly focus

H. B. Larsen (✉)
Department of Geography, University of Bergen, Bergen, Norway
e-mail: heidi.bjoennes@gmail.com

K. B. Lindkvist, T. Trondsen (eds.), *Nordic-Iberian Cod Value Chains*,
MARE Publication Series 8, DOI 10.1007/978-3-319-16405-2_6

on a 'green' approach to seafood in the Western world has especially emphasized health issues and eco-labeling (Brécard et al. 2009; Salladarré et al. 2010). This chapter will discuss the quality turn, through the concept of conventions, focusing on what will henceforth be referred to as green, civic and domestic conventions.

During the last 20 years there has been a restructuring in the Spanish market concerning the quality of salted fish products (Lindkvist et al. 2008). Salted cod (20–23 % salted content), or bacalao in Spanish, has over centuries had a special place in Spanish food tradition and culture (see Chap. 5). The present most prominent indicator of the quality of salted cod in Spain is a 'nuclear' white color, which is obtained through better production, shorter cure for the fish or phosphate injection. Phosphate increases weight, it retains the water in the fish and increases the whiteness of the fish meat (Esaissen et al. 2005). Lindkvist et al. (2008) were among the first scholars to address the addition of phosphate to salted fish sold in the Spanish market. The traditional conservation process, where salt was the only additive, gave the fish a slightly golden color. The new production processes give a 'nuclear' white color. Given the central position of salted cod in Spanish cuisine and the love Spanish consumers have for this traditional product, I initially took for granted that phosphate was a passing trend and that both producers and consumers would soon come to their senses.

After extensive research in the Spanish salted fish market, I am less confident. Hence, I ask, are consumers and producers in Spain approaching a time when they will demand a more natural color for salted fish and resist the addition of artificial additives? The question is inspired both by the 'quality turn' (especially the increased focus on healthy, clean and sustainable food products) and by a significant disagreement among the Spanish producers themselves about what constitutes a good salted fish product. The increased focus on sustainability and clean food in Europe probably prompted the existing civic and green values, rules and conventions, and this in turn may change current public expectations about food products. Based on surveys of and interviews with Spanish salted fish producers, interviews with consumers, and interviews with restaurant (*sidreria*) owners, I will examine the possibility of an increase in natural white salted fish in the Spanish market. The next section will discuss the theoretical framework of this chapter. Sections 3 and 4 analyze regional changes in color preferences and how changes in conventions are taking place among the actors in the value chain. Section 5 summarizes.

6.2 Green Conventions in an Institutional Framework

Economic life takes place in markets. Reiffenstein and Hayter explain that, "markets are socially constructed in particular times and places and their operations reflect social and political as well as economic goals" (2006, p. 508). They maintain that economic geographers have focused on either transaction costs or embeddedness to understand the nature of markets, whereas they use both perspectives in their approach. The study of embeddedness in economic life has concentrated on analyzing the social aspects of economic deeds. Murdoch et al. (2000) argue that embeddedness should be extended to include natural as well as social relations. The food sector's embeddedness in nature and the focus on food and ecological condition will probably matter even more in the future.

Chapter 2 in this volume presents and discusses how geographic embeddedness in markets matter because conventions associated with a food product are linked to places, culture, traditions, recipes and history. Convention theory and orders of worth have increasingly been used as an analytical framework in geographic studies of the food sector and understandings of quality in food production (Murdoch and Miele 1999; Ponte and Gibbon 2005; Lindkvist and Sánchez 2008; Ponte 2009; Sánchez-Hernández 2011). When people come together in a market, they need to be able to distance themselves from their own personal opinions in order to reach agreements on specific properties of external goods of some sort that need to be defined in general terms.

The six "orders of worth" concept (market, domestic, industrial, civic, opinion and inspiration) as discussed in Chap. 2, originates from Boltanski and Thévenot and their approach to justification in the economic worlds based on a typology of general agreements. They developed six orders of worth and modes of justification based on higher common principles. Allaire (2010, p. 169) refers to Max Weber's definition of order from 1947: "an order ... can roughly be defined as a prescription for how to act that has acquired a certain independence in the minds of the individual actors." Based on these definitions of order, conventions are quite similar in content. A convention is characterized as a valid order because conventions are guaranteed, and they are sanctioned through disapproval if they are broken. Conventions can be seen as a collective framework, the accepted way in which we act in our daily lives and in the market sphere. The expression of conventions changes over time, but their role as framework remains.

The question of artificial additives in salted fish production may concern some tension between *production conventions* (efficiency and cost) and *consumption conventions* (quality and value) (see Chap. 2). The six kinds of justification (Boltanski and Thévenot 2006), are the basis for values and interests that guide consumer decisions. This chapter emphasizes the green and civic conventions in order to focus on the quality turn. 'Environmental conventions' (Murdoch et al. 2000) and 'green orders of worth' (Boltanski and Thévenot 1999; Thévenot et al. 2000) has a contemporary introduction. In their 1999 article, Boltanski and Thévenot suggested a green order of worth in addition to their original six orders of worth. Thévenot et al. (2000) elaborated the green order of worth by claiming that "'green' worth is gaining specificity but is still often used in combination with other types of justification" (Thévenot et al. 2000, p. 237). A green order of worth focuses on environmental friendliness, renewability and sustainability (Thévenot et al. 2000, p. 241). Murdoch et al. (2000) introduced the ecological convention to place nature at the center of food chain analysis. Ecological concerns can easily be placed in domestic conventions because a return to nature in many ways means a return to local and small-scale production (Murdoch et al. 2000, p. 115). They further suggest that the ecological convention notion can shift the existing pattern of production from one that is against nature to one that works with nature. A turn toward green values may affect the negotiation of quality.

As a category of merchandise, food stands out because consumption of food directly affects our health and increases our focus on civic conventions (see Chap. 2). Seafood from captive fisheries is the last primarily wild-caught food readily available in restaurants and supermarkets. Hence, seafood represents an intersection between culture, nature and green conventions. Mansfield discusses the understandings and implications of quality in seafood in her articles on the surimi industry (Mansfield 2003). Quality can function as an institution based on conventions (Allaire 2010). Sánchez-Hernández (2011) describes conventions as the basic procedures used and decisions made by producers to give their products the features that consumers prefer. The physical aspects of quality in food can be measured and standardized through industrial and market conventions. The social construct of quality is determined by conventions, preferences and values and is based on the relationship between producers, consumers, retailers and other actors in the value chain (Mansfield 2003). There is a continual negotiation among these actors about what constitutes quality in a given food product. Consumers' conventions of quality are influenced by values, beliefs, income and geographic context. The quality turn describes how consumers have added additional features such as health values and green values to the understanding of quality. When the quality of a given product is negotiated, the actors involved make their arguments and reach compromises about the different underlying values and conventions. Supermarkets go green to keep their image clean. Customers go green to preserve their health and the fish stocks for coming generations. Producers go green to ensure sustainability and to stay in business. Researchers have not examined green conventions with respect to how such conventions can affect quality and conventional value chains. However here it is worth mentioning an organization securing ecological conventions through certifying sustainable fisheries worldwide, the Marine Stewardship Council (MSC). Currently 212 fisheries worldwide are certified according to MSC's environmental standards (www.msc.org).

The discussion of quality in this paper focuses on the color of fish flesh, although a number of other product features are important as well, such as texture, thickness and dryness. The focus will be on color as an expression of quality because color is essential to the consumers' definition of quality and because color is influenced by the production process and the treatment of fish throughout the value chain. The preference for white color can also be seen as somewhat contradictory because, although white may signify a clean, unpolluted fish, often the white color is created by adding chemicals. Qualities are geographically embedded and color preferences are an expression of regional conventions.

6.3 From Yellow to White: the Regional Embeddedness of Color Preferences

Value chains are connected to spaces and places and they combine conventions and natural, social and technological resources (Murdoch and Miele 2004). As a final market, Spain can be broken down into regional markets, each with its own conventions and quality preferences (see Chap. 5). The salted fish market in Spain has changed since the 1990s from a traditional market with few and known processes to

Fig. 6.1 Regional color preferences for salted cod in 1990

spatially different markets with heterogeneous products and production processes. Over the same period, however, color preferences in the Spanish salted fish market have become homogeneous. In 2011, I conducted structured telephone interviews with 88 % of the ANFABASA [1] members. Eighty-four percent of the Spanish salted fish producers surveyed agreed that white color is important to Spanish consumers.

Figure 6.1 shows the geographically embedded quality preferences for the color of salted cod in 1990. Twenty years ago, regional preferences were diverse but still slightly favored a preference for a golden fish color. This quality preference was based on tradition and domestic conventions. If we jump ahead 20 years, the situation has changed and there is a clear preference for white fish, as shown in Fig. 6.2.

[1] The Asociación Nacional de Fabricantes de Bacalao y Salazones was established in 2008 and includes 80 % of the Spanish producers of salted fish.

Fig. 6.2 Regional color preferences for salted cod in 2010

> Before, we used the dried salted cod, the one that was prepared to be conserved for a long time, but we don't use this type anymore in restaurants or gastronomic societies. Nowadays the dominant 'middle treated or half-cured' cod is preferred; it is not very salted. But you can still find the old type. The new ways are better. (Director of a cooking school, Basque Country)

Since the mid-1990s, the injected brine solution has frequently been supplemented with phosphates and antioxidants (Lindkvist et al. 2008). Modern processing gives the producers a higher yield because the fish retains a higher fluid content as well as a whiter color.

> France and Iceland have created the market for white fish. (Spanish producer)

The modern version of the artificially white color regime in salted cod was introduced by Iceland, when they established a market in Catalonia, a region with historical preferences for white fish (see Chap. 3). Complaints from Catalonia concerning the yellow Norwegian fish were registered 100 years ago (see Chap. 3). From Catalonia, the preference for white color spread throughout Spain. There are many reasons why white color now is perceived as good quality in the Spanish market. One reason is the market strategies of the Icelandic producers (Lindkvist et al. 2008), but the conventions or product expectations of the Spanish people have influenced this preference as well. Consumers in Barcelona states that white color in fish indicates high quality, as white signify cleanliness (Lindkvist and Røkenes 2009). In the wealthier regions, such as the Basque Country, Navarre, Argon and the larger cities such as Madrid and Barcelona, the convention for whiteness is dominant. The fresh fish market constitutes the largest part of the overall seafood market in Spain. The white color of fresh fish indicates health and social status, and consumers who prize civic values and the distinction of healthy food will pay higher prices for white fish (Sánchez-Hernández 2011). However, there are regional differences in quality demands. The traditional, natural-colored salted fish is still appreciated in several regions and it plays an important role in re-exports (see Chap. 7). In many regions, the demand for salted cod is driven by the desire for traditional dishes (Gallart et al. 2005). The traditional more yellowish fish is cherished in regions such as Andalucía and the Levante (see Chap. 5), and *tipo ingles*[2] is still popular in the area around Valencia. However, even there the white quality convention is gaining popularity. So, if the consumers are happy with 'nuclear' white fish, why is it likely that we will see a change in the market toward a more natural-colored fish?

6.4 Toward a More Natural Salted Cod

The current popular white quality might be more according to producer conventions, efficiency and cost control than the consumer conventions, quality and value. Injected brine mixed with phosphate increases the fish's water content and hence its saleable weight. In my quest to investigate the possibility of a change toward a more natural white fish, I will examine conventions of the supply, demand and regulations in the value chain for Spanish salted cod.

a. Supply side

In my investigation of the ANFABASA members' conventions or rules about product properties, 92 % of the producers interviewed agreed that the consumer preference for white color has changed the production process for salted cod. Phosphate, which is one reason for the white color, has been the most important factor in restructuring of the Spanish salted fish market. It has changed the industry's conventions about the most efficient way to produce salted cod. Catalonian producers have been a catalyst for the white color. They have clearly stated that they expect the Catalonian conventions of white, thick and juicy fish to spread to the rest of Spain (see Chap. 5). Despite the increased preference for white color, there is discord between the regional

[2] Cod that has been salted, washed and then salted one more time (see Chap. 5) Espinosa and Martínez, this volume.)

Table 6.1 ANFABASA members' reactions to arguments concerning civic and green conventions

Argument	Agree/partly agree	Means nothing	Disagree/partly disagree
If consumers and distributors in Spain knew that salted fish products could contain chemicals such as phosphate, they would choose products free from chemicals	25	8	5
Spanish salted fish producers are concerned about the sustainability of fisheries and environmental issues	30	2	6
Total number interviewed 38			

markets, as some regions are still guided by domestic conventions and favor the traditional product (see Fig. 6.2). A number of producers commented that their region prefers the more golden-colored fish. Despite the increasing preference for white fish and the potential extra profit in selling water, several Spanish producers are skeptical about adding chemicals such as phosphate to the brine to obtain the whiteness of the product and prefer the traditionally produced fish.

> We are very traditional and not interested in developing new products. (Spanish producer)

Spanish salted fish producers established ANFABASA to promote salted fish and to coordinate the industry. ANFABASA focuses on the quality of raw material, technical efficiency and a more coherent and consumer-friendly labeling. The product qualities that ANFABASA emphasizes are healthfulness, tradition, recipe versatility, storage life, origin in wild fisheries and continuous supply (Suescun 2008). Producers of salted fish are somewhat conflicted about their product, as they need to ensure profitability, but at the same time they are promoting a product with long traditions. In addition, Spanish producers are increasingly concerned about the environment and sustainable fisheries, as shown in Table 6.1.

Players vertically upstream in the production chain, such as Norwegian salted fish producers, are aware of this and promote their recently acquired Marine Stewardship Council (MSC) certificate for their cod fisheries when they approach the Spanish market. These conflicting considerations require the Spanish salted fish producers to navigate between industrial, domestic, market, opinion (trends), civic and green conventions. Given the possibility of increased profit by using artificial means to whiten fish, as well as signals from the market concerning preferences for natural white fish, what does an investigation of the demand side reveal?

b. Demand side

As mentioned in the theory section, quality is negotiated and the producers try to produce fish with the quality features that their consumers prefer. To see a change toward natural white fish, there must be a demand among the consumers. If we look at final consumers, restaurants and supermarkets, what is the attitude toward green and civic conventions? In Europe, a turn toward civic and green conventions can be seen by consumers' willingness to pay for green products (Brécard et al. 2009). Spain has great potential to follow this trend. In 2006, only 0.7 % of total food spending in Spain was devoted to organic products, largely because domestic organic products were sold to foreign markets (Gil and Soler 2006). However, enlargement of the domestic market could be crucial for Spanish producers because of increased competition in the European market. In their investigation of Spanish consumers' willingness to pay for organic olive oil, Gil and Soler (2006) found that buying decisions were mainly motivated by health concerns. Organic products are thought to be healthier because they are more natural. These researchers concluded that better information would make organic markets grow because they found a positive relationship between knowledge and willingness to pay (Gil and Soler 2006, p. 122). Spanish consumers associate the word 'organic' with being ecological, natural and not environmentally damaging (Aarset et al. 2004). Organic food should taste better and the organic label should be an indicator of quality (Aarset et al. 2004). The willingness to pay for 'green' products is also motivated by the consumer's desire to be regarded as socially responsible (Salladarré et al. 2010). According to Thøgersen (2010), organic food products are considered luxury goods and are more sensitive to economic upturns and downturns than everyday necessities. Younger consumer groups showed greater concern for the production process and the origin of products. Older consumer groups had better knowledge of recipes. Younger generations are more sensitive to environmental issues (Salladarré et al. 2010). Consumers appear to look at ecolabeling as a way to certify the production process (Salladarré et al. 2010). When consumers do not trust the fishing industry and its ability to regulate itself, the demand for ecolabels might be higher (Salladarré et al. 2010). Several scholars have shown the importance of knowledge about fish stocks, production processes and environmental consequences for understanding green consumer practices and conventions (Gil and Soler 2010; Brécard et al. 2009; Salladarré et al. 2010).

> We don't know how they treat the cod. We are consumers. We are not experts in the process. (Members of Sociedad Gastronomica Basque Country)

Currently, most Spanish consumers do not know that chemicals are added to salted cod to secure its white color and, as indirectly confirmed by Table 6.1, they would probably choose otherwise if they had the correct information. Other reports also indicate that the consumers will most likely change their attitudes when they learn about the added chemicals in salted cod (Bakke 2010; Santana 2010). Consumers are becoming more aware. Workers in supermarkets and fish shops have found that customers are now more likely to ask for natural products and demand information about the products they are buying (Bakke 2010). Changes in conventions create new rules for individual action. According to Seafood Norway's representative in Madrid, consumers presented with thick, nuclear, white salted cod (with added phosphate) and a slimmer, natural golden piece, prefer the former—if the choice is based on visual impression alone. However, when given information about how the fish was processed, they prefer the golden fish. Despite the economic hardships in Spain, consumers choose products that give the most 'health for the money' (Olsen and Johansen 2011). Health and civic values are gaining ground in Spain. Consumers

Table 6.2 *Sidreria* owners' quality conventions and knowledge about additives

Sidreria	Origin of distributor	Origin of cod	How much cod do you use per week in high season?	Color preference?	Do you know phosphate can be added to the fish?	Do you desalt the cod yourself?
(A)	Basque country	Iceland	300 kg	White	No	No
(B)	Basque country	Don't know	50 kg	White	No	Yes
(C)	Basque country	Norway	400 kg	Yellow	Yes	No
(D)	Basque country	Iceland	30 kg	White–yellowish	No	No
(E)	Basque country	Iceland	80–100 kg	Nearly white	No	No
(F)	Basque country	Faroe Islands	100 kg	White	No	Yes
(G)	Basque country	Iceland or Norway	20 kg	White	No	No

prefer the white, thick fish based on the belief that what is good for the eyes is good for the body (Lindkvist et al. 2008). Bakke (2010) found that it was essential to salted cod consumers that the fish was healthy, wild and captured in cold clean waters, indicating a preference for something natural and genuine. However, once more there are contradictions. Some Spanish producers say they are willing to put phosphate on the label and they claim that this would not affect consumers' preferences (personal conversation, Brussels, May 2011).

I interviewed owners of *sidrerias* (cider houses) in the Basque region during field studies in 2010 concerning their quality conventions and their views on additives. Salted cod is an important part of the traditional menu in *sidrerias,* and people of all ages consume salted cod there.

Table 6.2 indicates preferences for a white salted cod among the *sidreria* owners. White is the dominant quality feature, but there are those who prefer a touch of gold. The table also demonstrates the lack of knowledge about additives. Even though the cider house owners are skilled purveyors of salted cod, with hundreds of kilograms of fish passing through their restaurant every week, they have little knowledge of the production process. They entrust their suppliers with control over the product. In earlier times, the *sidreria* owners would desalt the fish themselves, but now they trust the producer with this work, effectively handing over control of the product to them.

Market values and quality preferences are also communicated to producers through retailers. Many regions in Spain are characterized by a traditional shopping pattern that relies on corner shops, market stalls and specialized grocery outlets. Even though the so-called supermarket revolution (Ritzer 2010) took place in Spain decades ago (El Corte Ingels—1940, Carrefour—1957, Eroski—1969, Mercadona—1977), supermarkets are still affecting the Spanish shopping pattern. In Madrid, people are more likely to make one stop for food purchases than to visit multiple shops (see Chap. 5). Shoppers at super- and hypermarkets are influential users because of the volume they buy. They know the products and they expect economic benefits from solutions that meet their needs. The large retailers may hold a little more power than the traditional channels.

> Supermarkets set the norms. The company adjusts to them. (Spanish producer)

When conventional supermarket chains focus on organic food or certifications, it increases the consumer's interest and knowledge of these products (Thøgersen 2010). In France, two major supermarket chains (Carrefour and Intermarché) use eco-labels to guarantee that their fish come from sustainable fisheries (Salladarré et al. 2010). Leading supermarkets such as Carrefour are concerned with sustainability and complying with regulations : "Sustainable development is a key part of the company's strategy" (www.carrefour.com). For the second year in a row, Carrefour is hosting an annual MSC campaign[3] in France (SeafoodSource.com). The Norwegian Seafood Council has assisted in the negotiation of a deal between Carrefour and some Norwegian salted fish producers. According to the Norwegian Seafood Council's representative in Madrid, Carrefour wishes to change to Norwegian salted fish in its Spanish shops because of the sustainability[4] of the Norwegian cod fisheries, the cleanliness and lack of chemical additives (because of strict Norwegian legislation) in their products, and the availability of funds to market the Norwegian fish. Other major Spanish supermarket chains have clearly stated that they want phosphate-injected fish because the consumers want thick white fish (presentation from a representative of one of the leading Spanish supermarket chains in Santiago de Compostela 2010). Once more, there are contradictions between the actors' conventions in the value chain. Maybe laws and regulations can coordinate conventions in the value chain more effectively.

c. Regulation

Additives in food are controlled through formal conventions such as laws and regulations. In the EU, all kinds of phosphates are banned in salted fish under Directive 95/2/EC on food additives other than colors and sweeteners. Although they are controversial, phosphates are used in the seafood processing industry to improve product quality. Added phosphates slow microorganism growth, reduce drip-loss on thawing, maintain product flavor and muscle structure, and add weight (Neuacher 2008). Although phosphate is naturally present in a number of foods, there is research claiming that added phosphate may accelerate signs of aging (Ohnishi and Razzague 2010), lead to kidney problems (Gutiérrez et al. 2010), and may be associated with ADHD

---

[3] Les Jours Bleus, an annual in-store campaign to promote sustainable seafood, in cooperation with the Marine Stewardship Council (MSC), and Findus and Labeyrie.

[4] Norwegian cod and haddock fisheries are MSC certified outside the twelve nautical mile border.

(www.phosadd.com). The Food and Agriculture Organization of the United Nations (www.fao.org) states that:

> The first and universal effect of all polyphosphate treatment is to increase the weight of the fish by retaining water. The weight gain is of no technological benefit, and represents to the producer a gain in weight of product sold. This weight gain, which amounts to selling added water, clearly must not be the aim of polyphosphate treatment; polyphosphates should be added to fish only for technically justifiable purposes. Polyphosphates cannot significantly improve the eating quality of fish, although claims in this respect are often made.

For a time it seemed that phosphate was on the way out of salted fish production. In September 2008, the European Free Trade Association (EFTA) Surveillance Authority (hereinafter ESA) opened a case against Iceland for breach of Directive 95/2/EC. The Icelandic Food and Veterinary Authority (hereinafter MAST) informed ESA in April 2009 that it had decided to reform the fishing industry in Iceland and that the use of polyphosphates in the production of salted fish was prohibited. On this notice, ESA closed the case in November 2009. In September 2010, representatives from MAST stated on Icelandic National Broadcasting that MAST would not enforce the polyphosphate ban before January 2011 because MAST had applied to the EU Commission for permission to use polyphosphates in salted fish. The ESA responded with a "letter of formal notice to Iceland for failure to enforce the ban on polyphosphates in salted fish" (event No. 578736) in December 2010, giving Iceland 10 days to respond. MAST then sent a letter to all salted fish producers in Iceland in December 2010 ordering them to abide by the rules banning phosphates in salted fish production.

Based on the ESA's actions, one would assume that this was the end of phosphates in the production of salted fish. On the contrary, however, several producers have approached the EU with requests to use diphosphates (E450), triphosphates (E451) and polyphosphates (E452) and the answer was, "The producers will be free to choose whether or not they will use phosphates in wet salted fish" (E-002097/2012, answer to a written question). Therefore, the road ahead is still uncertain concerning the ban on phosphates.

6.5 Discussion and Conclusion

This chapter has examined the possibility of an increase in natural white salted fish in the Spanish market. In the introduction, I asked whether consumers and producers in Spain would demand a more natural color for salted fish and resist the addition of artificial additives. Looking at the evidence presented in this paper, we can see that there are conflicting conventions in the links in the value chain and there is not a clear answer. In an attempt to reach a conclusion, I will first summarize the arguments for a change toward a more natural fish color and then arguments for a continuation of the present practices.

There are several indications that a natural white fish may increase its market share. The first indication is ANFABASA, an organization promoting domestic, civic and green conventions. Second, the regions and producers that still prefer the traditional way of production and a natural fish also suggest the potential for a natural white fish market. Although a preference for white color (as an indicator of quality) is clear in most regions of Spain, as shown in Fig. 6.2, there are still those who prefer the golden, traditional product. The third indication is the Spanish consumers' increased focus on health and green values and their willingness to pay for these values. Spanish consumers are willing to pay for healthy food, but currently they do not have sufficient information about the healthfulness of their food products to make informed decisions. The fourth indication is the super- and hypermarkets' increased focus on organic products, third-party labeling and sustainability. However, information is crucial to strengthening civic, green and domestic conventions, as it is these conventions that favor natural white salted fish.

In opposition to these indications are strong arguments that the current practices will continue. Fish producers that use phosphates increase their profits by selling water and reducing production time. Consumers have become used to snow-white fish with lower salt content, and the white color has been established as the central quality criterion. Catalonia has been the catalyst for the new white fish and actors in the value chain there clearly want this trend to continue. Industrial, market and public conventions support the current practice. For a value chain to be successful, its conventions must be coordinated. The value chain between Iceland and Spain has been coordinated according to illegal practices driven by industrial conventions. The coordination around an illegal additive has restructured the market and seems to be permanent. Given that the EU now opens for phosphates in salted fish, the producers will be instructed to put all the ingredients used on the label. Hence, the consumers will have the information they need to make an informed choice.

Allaire (2010) refers to Hirschman, who introduced the concepts of 'exit' and 'voice.' When customers are confronted with a decline in quality, they have two options, exit or voice. Those who exit look for alternatives, whereas voicers participate in an alternative construction. As shown by the example with the cider house owners, Spanish consumers have little knowledge of the production process and have given control to the producers. This leaves the industry with a dilemma. If the industry educates people about production practices, this may raise concerns among consumers, but failing to educate people may lead to greater problems in the future. The Norwegian salted fish industry is afraid that consumers will exit the salted fish market altogether if there is a focus on the addition of phosphates.

Norway's approach to the Spanish market has been described throughout the chapter. Spanish salted fish producers need a constant supply of fish. In the near future Spanish producers might be interested in increasing their cooperation with countries that have abundant and sustainable fisheries. Norwegian fishing regulations have been successful in past decades, leaving the Norwegian fishing industry with a substantial volume of products to offer the international markets. For this reason, Spanish producers would benefit from a relationship with Norwegian producers and their more naturally colored fish, thereby broadening their choice of suppliers. Independent of the phosphate issue, Norway needs to change its current focus from 'volume and good enough quality' to one of 'good and right quality', to be able to reenter the Spanish domestic market.

Thus, the conclusion seems to be that there will be no turn to natural white fish in the near future, as producer conventions and the existing industrial and market conventions are too strong; however, there will still be room for the traditional product. Information may be a way to strengthen the position of the traditional product. Information that focuses on cold, clean water, sustainability, traceability and cleanliness, in accordance with consumer conventions, can allow producers to differentiate their products and reach younger consumers. As these comments suggest, it is not necessary to denigrate other products to promote one's own.

References

Aarset BS, Beckmann S, Bigne E, Beveridge M, Bjorndal T, Bunting J, McDonagh P, Mariojouls C, Muir J, Prothero A, Reisch L, Smith A, Tveteras R, Young J (2004) The European consumers' understanding and perception of the 'organic' food regime: the case of aquaculture. Brit Food J 106(2):93–105

Allaire G (2010) Applying economic sociology to understand the meaning of "quality" in food markets. Int Assoc Agric Econ 41:167–180

Bakke M-BD (2010) Spanske konvensjoner og bygging av merkevare for norsk saltfisk. In: The Norwegian-Spanish salted fish project. University of Bergen, Bergen, (02/10)

Boltanski L, Thévenot L (1999) The sociology of critical capacity. Eur JSoc Theory 2(3):359–377

Boltanski L, Thévenot L (2006/1991) On justification economies of worth. Princeton University Press, Princeton

Brécard D, Hlaimi B, Lucas S, Perraudeau Y, Salladarré F (2009) Determinants of demand for green products: an application to eco-label demand for fish in Europe. Ecol Econ 69:115–125

Esaissen M, Østli J, Joensen S, Prytz K, Olsen JV, Carlehög M, Elvevoll E, Richardsen R (2005) Brining of cod fillets: effects of phosphate, salt, glucose, ascorbate and starch on yield, sensory quality and consumers liking. LWT—Food Sci Technol 38:641–649

Gallart L J, Heidie M, Østli J (2005) Market situation of Norwegian Bacalao on the Mediterranean Spanish coast. Økonomisk Fiskeriforskning, 15:51–55

Gil JM, Soler F (2006) Knowledge and willingness to pay for organic food in Spain: evidence from experimental auctions. Food Econ Acta Agric Scand C 3:109–124

Gutiérrez OM, Anderson C, Isakova T, Scialla J, Negrea L, Anderson AH, Bellovich K, Chen J, Robinson N, Ojo A, Lash J, Feldman HI, Wolf M (2010) Low socioeconomic status associates with higher serum phosphate irrespective of race. J Am Soc Nephrol 21(11):1953–1960

Lindkvist KB, Røkenes A (2009) Med saltfisk på menyen The Norwegian–Spanish Salted Fish Project. University of Bergen, Bergen 01/09 Working paper

Lindkvist KB, Sánchez JL (2008) Conventions and innovations: a comparison of two localized natural resource-based industries. Reg Stud 42(3):343–354

Lindkvist KB, Gallart-Jornet L, Stabell MC (2008) The restructuring of the Spanish salted fish market. Can Geogr 52(1):105–120

Mansfield B (2003) Spatializing globalization: a "geography of quality" in the seafood industry. Econ Geogr 79:1–16

Murdoch J, Miele M (1999) Back to nature": changing "worlds of production" in the food sector. Soc Rural 39:465–483

Murdoch J, Miele M (2004) Culinary networks and cultural connections: a conventions perspective. In: Amin A, Trift N (eds) The blackwell cultural economy reader. Blackwell, Oxford, 231–248

Murdoch J, Marsden T, Banks J (2000) Quality, nature and embeddedness: some theoretical considerations in the context of the food sector. Econ Geogr 76(2):107–125

Neuacher H (2008) The use and abuse of polyphosphates, Intrafish. http://www.intrafish.com/sfp/news/article1245006.ece. Accessed 21 April 2015. (published, 21.08.2008)

Ohnishi M, Razzague S (2010) Dietary and genetic evidence for phosphate toxicity accelerating mammalian aging. FASEB J 24(9):3562–3571
Olsen K, Johansen O (2011) Norske torskefisker markedsplan 2010–2012, Norwegian Seafood Council. Tromsø
Ponte S (2009) Governing through quality: conventions and supply relations in the value chain for South African wine. Sociol Ruralis 49(3):236–257
Ponte S, Gibbon P (2005) Quality standards, conventions and the governance of global value chains. Econ Soc 34:1–31.
Reiffenstein T, Hayter R (2006) Domestic timber auctions and flexibly specialized forestry in Japan. Can Geogr 50(4):503–525
Ritzer G (2010) Globalization: a basic text. West Sussex, Wiley-Blackwell
Salladarré F, Guillotreau P, Perraudeau Y, Monfort M-C (2010) The demand for seafood eco-labels in France. J Agric Food Ind Organ 8(10):1–23
Sánchez-Hernández JL (2011) The food value chain as a locus for (dis)agreement: conventions and qualities in the Spanish wine and Norwegian salted-cod industries. Geogr Ann Ser B 93(2):105–119
Santana L (2010) Konvensjonspåvirkning og bygging av merkevarer for norske saltfiskprodukter i Spania. The Norwegian–Spanish salted fish project. University of Bergen, Bergen, (03/10)
Suescun R (2008) Strategic Plan of ANFABASA, Presentation
Thévenot L, Moody M, Lafaye C (2000) Forms of valuing nature: arguments and modes of justification in French and American environmental disputes. In: Lamont M, Thévenot L (eds)Rethinking comparative cultural sociology. Cambridge University Press, Cambridge, pp 229–272
Thøgersen J (2010) Country differences in sustainable consumption: the case of organic food." J Macromarketing 30(2):171–185

Chapter 7
Conventions and Value-Chain Development in the Norwegian–Spanish Seafood Trade: The Case of Salted Fish

Knut Bjørn Lindkvist

Abstract This chapter focuses on the use of the convention concepts related to the model of value chains for salted fish production between Norway and Spain. The intention is twofold: to see how conventions affect value chains and to obtain new insights into the international seafood trade and the forces that influence the exporting of salted fish from Norway to Spain. Empirically, the presentation is based on investigations of the Norwegian perceptions of conventions that are governing value chain decisions.

Keywords Location and conventions · Conventions and value-chain organization · Norwegian perceptions of Spanish conventions · Norwegian strategies · Market obstacles

7.1 Introduction and Scope

When Norwegian producers experienced a reduction in their share of the Spanish salted fish market from nearly 40 to approximately 10 % during the 1990s, the general expectation was that the Norwegian industry would attempt to recapture its lost market share. This has not taken place and some reasons for this situation will be discussed here. The question is asked whether the Norwegian producers know the conventions that govern the Spanish market for salted fish production and how they eventually assess the possibilities to comply with these conventions.

This chapter claims that the reason for the lack of success in Spain lies in the Norwegian production and governance structures and institutions, and especially the conventions designed for markets other than the Spanish consumer market. Even if the Norwegian producers know the Spanish conventions, conditions may exist in the Norwegian production system or in Spain that prevent the development of production

K. B. Lindkvist (✉)
Department of Geography, University of Bergen, Bergen, Norway
e-mail: Knut.Lindkvist@geog.uib.no

Finnmark University College, Finnmark, Norway

K. B. Lindkvist, T. Trondsen (eds.), *Nordic-Iberian Cod Value Chains*,
MARE Publication Series 8, DOI 10.1007/978-3-319-16405-2_7

conventions in Norway from matching the Spanish market conventions. Central to the chapter will be the conclusions drawn by the Norwegians from their perceptions of the Spanish production and market conventions. The chapter will conclude on the practical consequences of their reading of the situation in the Spanish salted fish market.

The contribution of this chapter to the conventions debate is to reveal the importance of spatial and functional localization for participants in understanding conventions and product adaptation throughout the entire value chain.

The remainder of the paper is organized as follows. Section 7.2 discusses the role of conventions and value-chain organizations. Section 7.3 presents the methodology we employ, while Sect. 4 considers the Norwegian–Spanish value chain in salted fish. Section 7.5 examines the structures and process at play in this market and Section 6 introduces the role of the conventions and strategies employed. Section 7.7 examines some of the obstacles to these developments in consumer markets, and some concluding remarks are presented in the final section.

7.2 Conventions and Value-Chain Organization

The concept of the value chain and conventions linked to the food trade form the fundamental basis of the theoretical model (Fig. 2.2, Chap. 2) that functions as the analytical tool for the analysis of the lack of market adaptation by the Norwegian salted fish industry.

Chapter 2 referred to the value chain as a number of linked transactions, where the value adding process forwarded products in one direction and with payment in return. The flow of products and payments are as mentioned unfolded and coordinated within a context of influences, technological social and economic pressures. In the value chain, such pressures are exercised in both directions, from raw material producers to actors in the market, from market participants to primary producers. If market participants have product alternatives, the ability to comply with the conventions in the market may decide whether the primary producers have success in the markets or not. A fundamental question is therefore whether the producers in the value chain's first parts know the market conventions, and if so, how they perceive their ability to comply with the requirements as expressed in these conventions.

One way of controlling for insecurity of coordination is to establish institutions that work to secure the interests of firms under volatile competitive conditions (Cornish 1997; Gereffi et al. 2005). To learn to read the market conventions may often be one way to secure positive market conditions. Such strategic adaptation can be achieved through direct establishments, which are as close as possible to the end market, to the internalization of sales functions within the producing company itself, in which some employees work with sales contracts and market intelligence. However, as almost all Norwegian fish producers are without any permanent representation in the markets, firms with or without specific sales and intelligence functions also depend on good relations with all possible importers. Even more, the producers without representation in the market dependent on legal and social frameworks which allow them equal opportunities to conduct business processes without too many discriminatory obstacles in the value chain.

The social activities and relationships and the formal and informal governance rules (conventions) developed by the industry, function in reality as a substantial part of the framework that regulates and has a large influence on production processes (Gertler 1997; Dicken 1998). For example, Gertler (1997) raises the question of whether players will be able to create innovative and co-ordinated social relations, even if they wish to, if parts of this framework actually prohibit or otherwise prevent the relevant social relationships and the development of social capital. Figure 2.2 (Chap. 2) depicted some of these conventions as general parts of this regulation framework, but with an unequal effect on the chain units. In turn, these different outcomes arise from the composition of the regulatory arrangements of the different geographical and functional parts of the value chain.

Different social practices and differences in legislation will also create different regulating frameworks for the players in the chain. For instance, despite a strong common law for food in Europe, local practices will continue to influence local adaptation to general regulations. European practice for the use of polyphosphates in brine is one such example where practice is different from the law itself. (Lindkvist et al. 2008). In other fields, like the denial for the processing companies to own fishing vessels and dispose of quotas, the Norwegian industry is forced to compete on unequal terms with industrial competitors from other countries (Trondsen 2012). A spatial or geographical perspective is also important within each unit of a value chain and between the different chain units. Producers are often co-located in specific geographical areas. Over time, these actors will experience that their characteristic production institutions are influenced by the spatial proximity between them, and by the specific social practices and formal and informal rules (i.e. conventions) typical of the regions where they unfold (Boschma 2005; Crevoisier 2004). In some nature-based industries, such local institutions largely determine how specific parts of production and transactions should be performed (Lindkvist and Sanchez (2008). In such cases, this local influence is present, even if the authorities regulate all types of standardized (as well as local) food processing activities. However, exactly how much influence local conventions receive depends on the framework of the value chain.

The different regulatory contexts and socio economic practices into which any producer are embedded may influence how he interprets the conventions perceived in the market units of the value chain. This chapter will account for the producers' understanding of the market conventions, and how they view their own possibilities to comply with the market demands expressed by the market conventions.

If the conventions of the Norwegian commodity producer cannot change to fit consumer markets, the Norwegians will need to establish alternative value chains that will allow their conventions to unfold in the importing country. This will be possible through the existence of alternative value chains based on products and market conventions, which already find resonance among producers.

7.3 Methodology

The empirical material for this chapter is from several sources. The primary material is from a survey of 122 Norwegian producers involved in salted fish production in 2009. We interviewed 103 companies (a response rate of 84.4 %) using questionnaires originally sent by email. We also interviewed companies that did not initially respond via telephone. Another source we employed is the salted fish project (Espinosa and Martinez 2010), which involved intensive interviews conducted in all of the main Spanish regions, comprising 40 companies that produced salted fish, 50 fishmongers, several restaurants, and 50 individual consumers. In addition, we conducted an intensive field survey in autumn 2010 in Galicia in Spain, in which we interviewed key respondents about the state of the salted fish trade with Norway. We also interviewed four key informants in Spain and Portugal. Finally, we drew upon research from previous studies (Lindkvist and Sánchez 2008; Lindkvist et al. 2008). Note that some of the data sources referred to are the database of the Norwegian–Spanish Salted Fish Project[1].

7.4 The Norwegian–Spanish Value Chain in Operation

During the first decade of this millennium, only one Norwegian firm (Spanish owned) had established permanent production or sales facilities in Spain. Such avoidance of direct investments in the importing country was part of a traditional adaptation strategy among the Norwegians. Independent Norwegian fish brokers had instead organized the relationship to the buyers in Spain. They had developed their own social and economic networks with the Spanish using own relations with the Norwegian producers as their main social capital. They therefore provided the Spanish demand to the Norwegian producers.

Of the 33 Norwegian companies that supplied salted fish to Spain in 2009, only one again looked to Spain as its most important market, while 26 producers considered Spain as their second most important market (Table 7.1). Another six Norwegian producers concentrated on yet other markets other than Spain, with Spain as the number three market. This apparent lack of priority given to the Spanish market may suggest a first explanation why the Spaniards were generally unhappy with the quality, grade, and volume of their Norwegian purchases (Lindkvist et al. 2008). Spanish market conventions may not be of the most important concern for the Norwegian producers.

Table 7.1 also shows that 20 Norwegian producers for Spain (61 %) who also operate in other markets to which they give more priority, dispose of own export department. Without direct presence in Spain, almost all of them (except the Spanish

[1] NFT (Norwegian Research Association), project number 185126/I10.

Table 7.1 The market and functional structure of producers operating in the Spanish market

	Spain = 1 priority market	Spain = 2 priority market	Spain = 3 priority market	Total
With own export	1	15	4	20
Without own export	–	11	2	13
Total	1	26	6	33

Table 7.2 Market orientation among Norwegian producers

Norwegian producers' view	Spain as no. 1 market	Spain as no. 2 market	Spain as no. 3 market	Total
Spain as no. 1 market	1	–	–	1
Portugal as no. 1 market	–	21	2	23
Brazil as no. 1 market	–	–	2	2
Other countries as no. 1 market	–	5	2	7
Total	1	26	6	33

owned Norwegian company) may have fewer opportunities to gather market intelligence here. Their production range may be characterized by a focus on standard products that function as generic inputs to the processing units of other companies, in either Norway, Spain or other markets. This situation (Table 7.1) becomes even more obvious for the remaining 13 of the 33 Norwegian producers that do not even have a separate sales department, but are instead obliged to sell through the 20 remaining Norwegian producers with sales functions.

Table 7.2 shows that 23 producers operating in Spain target Portugal as their most important market, as many as 21of these firms target Spain as their number two market. Two of the "Portuguese" firms stated that Spain only is their third most important market (Table 7.5). Two firms focusing on Brazil as their most important market designated Spain as their third most important market. Finally, all of the seven additional firms working and focusing mainly in other countries consuming Norwegian salted fish referred to Spain as their second or third most important market.

Also Table 7.2 confirms again that the Spanish market is not a major concern for Norwegian producers. In fact, the volume of Norwegian salted fish exported to Spain decreased from 2007 to 2009 (Fig. 7.1), continuing a general downward trend beginning in 1998 (Lindkvist et al. 2008).

Lindkvist (2010) and Lindkvist and Sánchez (2008) have given several additional reasons for this decrease in the Norwegian share of the Spanish market. These include the fragmentation of the Norwegian value chain and the prohibition of the injection of phosphate into the salting brine. In addition, there is a lack of research focusing on export barriers among the production companies and the absence of a collaborative culture among such members of the production system, and this could

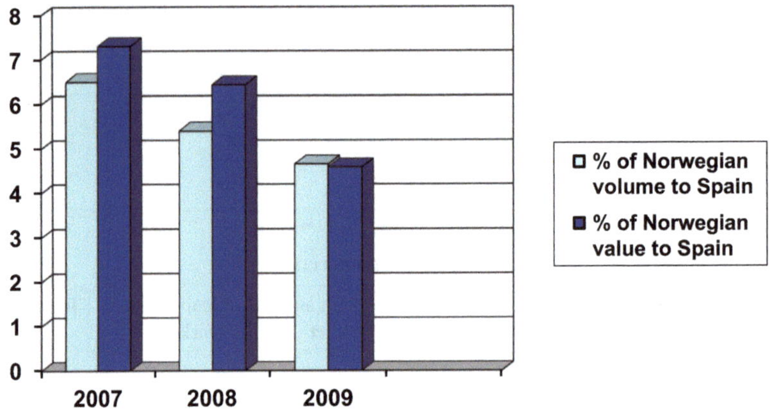

Fig. 7.1 Changes in Norwegian exports to Spain 2007–2009

have contributed to market failure. The overall outcome was a lack of innovation in and adaptation to the Spanish consumer market (Lindkvist 2010).

Table 7.2 indicates an additional explanation for this market outcome, but one related to that previously outlined. The almost total lack of a permanent presence in Spain of Norwegian manufacturers may indicate either a lack of knowledge of modern conventions in the Spanish salted fish industry or a lack of capacity for immediate movement within these markets. How well can Norwegian producers know the conventions of the Spanish market, and understand the finely tuned marked drivers when they are neither present in the market nor prioritize this market in the first place? To understand the choices of the Norwegian producers it is important to know the view that Norwegian producers after all have of the conventions governing production and trade in the Spanish salted fish market.

7.5 The Reasons for the Structure of the Norwegian–Spanish Value Chain: Norwegian Perceptions of Spanish Conventions and Norwegian Strategies[2]

We followed up the 2009 investigation using a study of the Norwegian producers and their perceptions of Spanish conventions. More than 84 % of the registered Norwegian exporters delivering to Spain participated in the study. We drew the conventions examined from Espinosa and Martinez (2010) and Sanchez-Hernández (2010), along with Drage Bakke (2010) and Santana (2010).

[2] A detailed overview on the different variables mentioned is found in Gonzalez-Lopez & al. (Chap. 8), Trondsen a & b (Chaps. 9 and 13, Xie & Myrland (Chap. 10), Xie (Chap. 11).

Table 7.3 Norwegian perceptions of Spanish market conventions and Norwegian domestic production conventions

Market preferences	Domestic production convention values
TASTE: Taste is an important quality factor to the consumers in Spain who buy salted fish	GOVERNANCE: Reliable governance rules in the production country (not Spain) that strengthen the product quality are important to the Spanish distributors when they decide which supplier will be chosen
COLOR: The white color of the bacalao (salted fish) is important to Spanish consumers	THE FISHING METHOD: The distributors in Spain think that the methods used in fishing influence the quality of the raw products for the production of salted fish
PRODUCTION ORIGIN: It is important for Spanish consumers to know which country the salted fish comes from	THE SUPPLIER QUALITY: Good supplies of quality products are important to the Spanish distributors when they decide which salted fish producers to do business with
THE PRICE–QUALITY relationship: Cheap salted fish signals poor product quality for the consumers in Spain	THE PRODUCER–PRODUCER RELATIONSHIP: Spanish producers and distributors have positive attitudes towards a better collaboration with Norwegian salted fish producers
THE DESALTED DEMAND: The consumers have increased their demands for desalted fish products because the products are easy to prepare and they save the consumers' time	THE RAW PRODUCT IMPORTANCE: Spanish producers think salted fish from fresh cod gives the best quality
THE PHOSPHATE: If Spanish consumers knew that salted fish products might be produced with the use of chemicals such as phosphates, they would certainly choose products where chemicals had not been used	
THE NORWAY SUPPLY: If we ask Spanish consumers today, they are confident that Norwegian suppliers will deliver salted fish products of a good quality	

Table 7.3 indicates what Norwegian producers *believed* Spanish producers and consumers were thinking about salted fish and their co-operation with the Norwegian industry. This Norwegian understanding of the Spanish preferences in the survey were categorized in two groups as either domestic production conventions or market preferences

The Spanish market preferences (Table 7.3) were perceived to be concerned with the preferred taste of Spanish consumers of the product, product color, country of origin of the raw product, the significance of the price for quality perception, how desalination trends are developing in the market, the market's acceptance of phosphates, and Spanish market confidence in Norwegian quality.

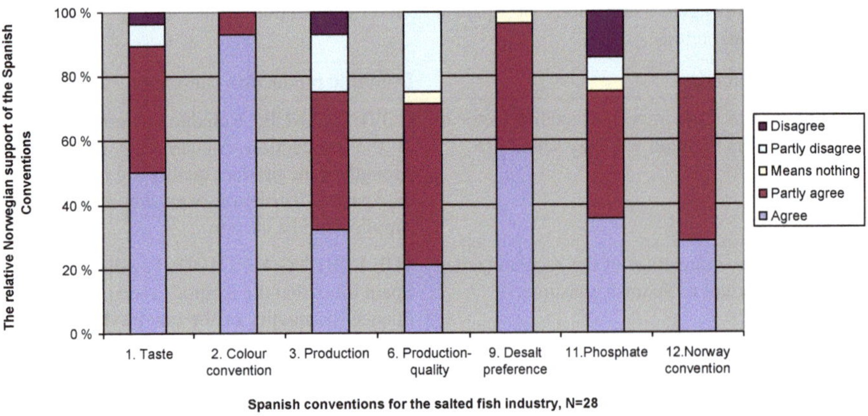

Fig. 7.2 Norwegian understanding of the Spanish market preferences

The Norwegians perceived how the Spanish domestic conventions (Table 7.3) concerned the governance rules in the value chain and their importance for Spanish producers. The Norwegians understood how the Spanish market favored specific fishing methods, what made the Spaniards consider suppliers to be delivery capable, when the Spanish producers considered a possible partnership with the Norwegians, and the type of raw material considered best for Spanish salted fish products.

Figure 7.2 shows reasonably strong agreement among Norwegian producers on which market preferences are applicable in the Spanish market for salted fish. The strongest agreement among Norwegian producers is about the dominance of the white color of salted fish: 93 % of Norwegian producers acknowledged the dominance of the white-flesh salted fish regime in Spain. The remaining producers partly agreed.

Second, 57 % of the Norwegians agreed that there is a desalination convention trend among Spanish consumers; 39 % partly agreed with the statement "consumers have increased their demands for desalted fish products because the products are easy to prepare and they save the consumer's time." The emphasis on taste as an important quality factor for consumers also has strong support among Norwegian suppliers to Spain. For example, 50 % strongly agreed with the statement "taste is an important quality factor to the consumers in Spain who buy salted fish" and also here 39 % partly agreed with this same statement.

There is somewhat greater disagreement in Norway about the remaining Spanish market preferences. The clearest disagreements among Norwegian producers are linked to product origin preferences (25 % disagree), phosphate injection (21 % disagree), the indication that a low price equals low quality (25 % disagree) or that the Spaniards have confidence in the quality of the Norwegian deliveries of salted fish (21 % disagree with the importance of this preference).

Figure 7.3 shows the same degree of common understanding among Norwegian producers of the Spanish production system and their domestic conventions. The greatest support found is for the understanding of the Spanish supplier convention

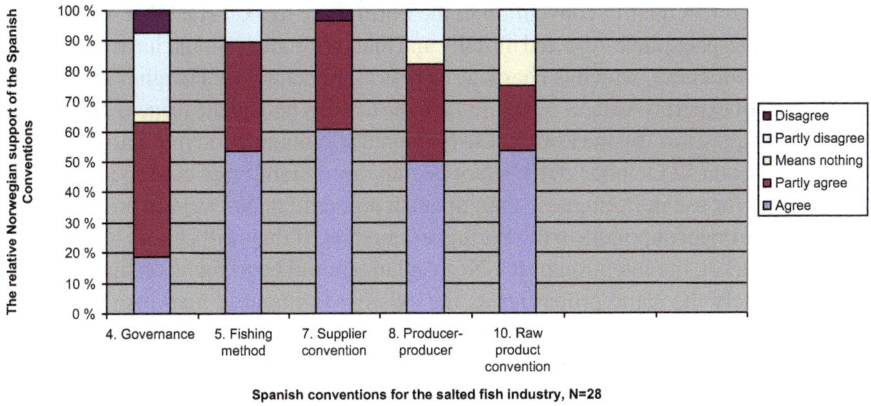

Fig. 7.3 Norwegian understanding of the Spanish domestic production attitudes

values saying "good supplies of quality products are important to the Spanish distributors when they decide which salted fish producers to do business with." Here, 61 % of the Norwegian producers strongly agreed with the statement; 36 % partly agreed. The fishing methods is considered by the Norwegians to be important for Spanish manufacturers: 54 % of Norwegian producers agreed with the statement that "the distributors in Spain think that the methods used in fishing, influence the quality of the raw products for the production of salted fish" and 36 % partly agreed.

There is also significant Norwegian agreement that the Spaniards exhibit a positive attitude to future collaboration with Norwegian producers, with 50 % of the Norwegians agreeing with this presumption and 32 % partly agreeing. The greatest disagreement among Norwegians concerns the convention that states "reliable governance rules in the production countries (not Spain) that strengthen the product quality are important to the Spanish distributors when they decide which supplier will be chosen." Here, 37 % do not support the existence of this convention value.

Nevertheless, there is generally strong agreement among Norwegian producers about the rules that apply to domestic production systems and markets in Spain. Further, it has not been possible to find any significant difference between Norwegian producers that have their own export departments and those that do not. The degree of support for the Spanish conventions applies to all types of Norwegian businesses regardless of location. Cross-linkages in Norway regarding ownership, the relatively small business environments, shared communication channels and similar types of Norwegian institutional networks (for example, through industrial organizations, conferences and information dissemination by the fish brokers) most likely contribute to the relative consensus among the Norwegians concerning the salted fish preferences that apply in the Spanish market.

Table 7.2 showed that Portugal was by far the more dominant market for the Norwegian producers that also aimed at Spain. About a quarter of all clipfish, and around half of all wet-salted fish went to Portugal (Seafood Norway).

The salted fish quality convention in the Portuguese market state that dried salted fish cannot be phosphate-injected initially and that the product should have a golden or yellowish whiteness, which is characterized as more natural by Haugnes (2010) and by a panel of experts (Anfaco 2010). In fact, the use of a phosphate brine is considered "out of the question" by the Portuguese (personal communication in Aveiro, Portugal to the author on 12 October 2010[3]). Norwegian raw materials are consequently more convenient for use in Portuguese than Spanish production. Norwegian producers are therefore stronger suppliers to the Portuguese market. If they still choose to sell to the Spanish market, the intention of the Norwegians should be to focus on markets with approximately the same properties as the existing Portuguese markets. This is not difficult to find even in Spain, as the Spanish salted fish markets are characterized as much more differentiated than the relatively uniform Portuguese market (Lindkvist et al. 2008). Accordingly, some Spanish markets exist with similar traits to the Portuguese ones.

Genuine Spanish producers themselves, however, have also maintained contact with the markets in Portugal and Brazil that still adhere to previous quality conventions (Espinosa and Martinez 2010). Spanish producers can actually work with all these markets using Norwegian fish as their raw material base.

The Norwegian exporters therefore know that the dominant Spanish conventions focusing on color, quality and the use of phosphates are irrelevant in "their" Spanish market. Those with the closest links to the importers in Spain know that they are producing for Portugal or for similar Spanish markets (markets for dried or semi-dried products, as for markets focusing wet salted fish with golden color)[4]. The Norwegians know that the Spanish markets still have many options available if they cannot meet the dominant Spanish demands and standards. Figure 7.4 indicates the consequences of this Norwegian market performance for the structure of the production chain established in the salted fish trade between Norway and Spain.

As shown in Fig. 7.4, the Norwegian–Spanish value chain obtains a two-armed or bifurcated form when it crosses the Spanish border. In reality, more than one value chain is effective. However, the idea of a bifurcated value chain opens for an immediate understanding of the mismatch between Norwegian and Spanish conventions. One branch of the chain is for internal consumption markets, including the H*o*tel/Restaurant/C*a*fé ("HORECA") market, supermarkets, and fishmongers, before it reaches final consumers. The other branch consists of the industry that uses Norwegian fish as a base to which it adds further value for foreign markets outside Spain, such as the Portuguese or Brazilian markets. When the Norwegian fish trade followed this trade path, it proves that the Spanish internal markets of modern products are of

---

[3] The information emerged in an intensive interview with the Secretary and a prominent producer of the association of salted fish producers in Aveiro.

[4] A Spanish producer in southern Spain claimed that during a 5-year period in the first decade after 2000 they bought 60–65 % of all Norwegian salted fish in Spain. This fish enters the Portuguese and Spanish markets where Norwegian qualities are demanded (a focus on dried fish, three-quarter dried, and golden wet salted). In Spain, Norwegian raw products are also used to some degree to supply cheaper products.

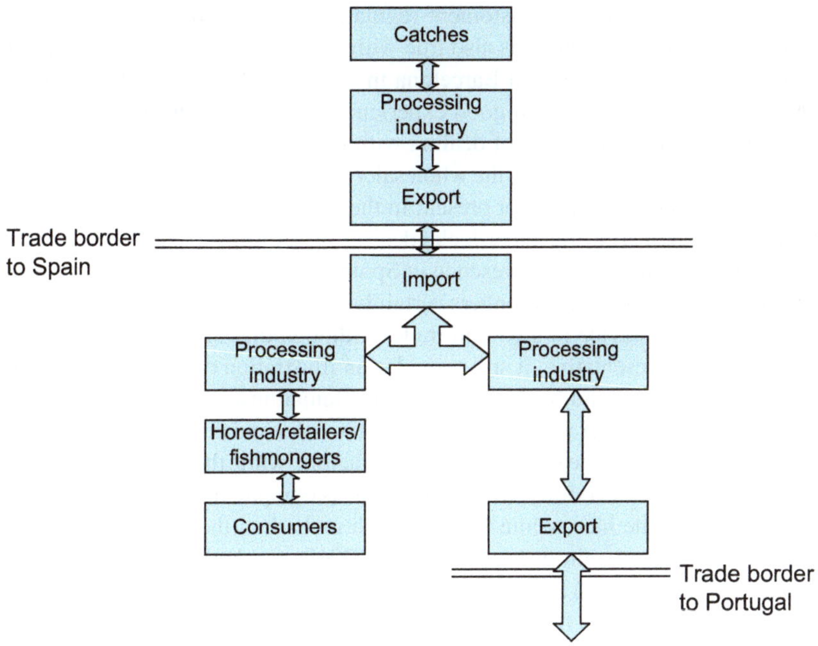

Fig. 7.4 The value chain for the Norwegian–Spanish salted fish trade with two Spanish branches

less interest and so less focused upon. This means in reality that the Norwegians have not involved themselves greatly with the Spanish consumer markets and thereby the modern consumer conventions focusing white fish products.

7.6 Obstacles to Entering the Horeca and Consumer Markets in Spain

Although important trade flows of salted fish from Norway to Spain go through the value chain only to be re-exported to other countries, the second arm of the chain (the HORCA and consumption branch) is also supplied from Norway. This is where Seafood Norway has focused its export efforts since autumn 2009. To what extent will the current Norwegian production and export structure, combined with the exporters' perceptions about their position in Spain, have an impact on opportunities to build up this part of the value chain?

Not only does the lack of integration between the units located in Norway characterize the structure of the Norwegian–Spanish chain. Apart from the single Spanish-owned Norwegian company, direct Norwegian activities end at the Spanish border. If we look to Iceland, the value chain of Iceland to Spain is better functionally integrated. The Icelandic salted fish producers are physically present in the Spanish market with their own representations.

Loyalty to suppliers and customers regulates the way Spanish producers and fishmongers trade their fish. This also true with customers, like Spanish restaurants, as shown by an investigation in Barcelona in spring 2009 (Lindkvist and Røkenes 2009). This loyalty stems from good experience over time with secure supplies of quality fish and accepted ways of dealing with minor disagreements about volume, quality and price. Consequently, the wholesaler is the decisive factor in this situation. As long as Norwegian fish is not present in the market, the wholesaler will rely on supplies from the Faroe Islands or Iceland.

In addition to the lack of a presence in Spain, the quality conventions expressed by, for example, Spanish fishmongers certainly have an influence when their clients seek assistance or advice in the retail trade. Although the following statement may not be wholly representative, it still provides an illustration of what appears to be a depressing state of affairs for Norwegian fish. Benjaminsen and Gabrielsen (2010) interviewed a fishmonger who considered that the color and thickness of the products displayed in the stalls were the most important factors, while the "real" product was of less interest. In this interview, Norwegian fish was compared to "a dry ball of *jamón serrano*", while "the nice, white Icelandic fish melted on the tongue, like a tender *jamón ibérico*" (Benjaminsen and Gabrielsen 2010, p. 14). Lindkvist et al. (2008) and later Espinosa (2011) have previously also reported this particular viewpoint of the quality of Norwegian fish.

Although some of the Spanish market and domestic conventions may sound a little depressing for Norwegian producers, who are not able to comply with several of them, other production conventions are no less depressing. Here, the well-known high Norwegian cost level of labor cost may be a barrier to consumer market entry. If Norwegian producers wish to enter the Spanish market with their own products processed in Norway, they may discover that their processing costs probably are a second barrier to success. Quite simply, it may not be possible to produce finished products with traditional manual methods for the Spanish market in Norway from a low-labor cost perspective, as the Norwegian level of salaries and social costs is many times higher than that in Spain (Benjaminsen and Gabrielsen 2010). The alternative for the Norwegian is obvious to substitute expensive labor with atomized technology.

Therefore, the competition with Icelandic fish in the arena for traditional production conventions is difficult to win in Spain (Trondsen 2012; Xie and Myrland 2010). According to Benjaminsen and Gabrielsen (2010), even the best Norwegian fish is in addition, unable to compete with Icelandic fish in terms of color and thickness. Therefore, Icelandic prices cannot be achieved by the Norwegians, "and we are therefore unable to reach for the prices achieved by the Icelanders for their products. If you also add the costs linked to value adding like cutting, fillet production and so on, this is clearly a hopeless arena to compete in. To increase profitability (company X) has to choose another strategy than further processing" (Benjaminsen and Gabrielsen 2010, p. 15).

Even if phosphates are not permitted in the processing of salted fish, either in the European Union (Anfaco 2010) or in Norway, according to respondents in Iceland, a great deal of the fish from Iceland and the Faroe Islands is phosphate injected

(Lindkvist et al. 2008). In addition, in the experience of Norwegian producers, both Spanish producers and consumers demand, accept and even desire phosphate production. This experience will probably dominate the perceptions of the Norwegians, while widespread skepticism in Spain about the use of phosphates is perhaps more difficult to detect without their presence in the market. A consequence of these and the other perceptions may be for Norwegian producers to stay away from the Spanish market; the conditions described here may also have contributed to the Norwegian avoidance until 2010 of the more demanding arm of the value chain that targets the consumer and HORECA markets.

7.7 Conclusion

This article has argued that the lack of success in the Spanish market is driven by Norwegian governance structures and institutional conventions designed for other markets. The discussion has shown that this assertion is at least partly correct. Certainly, the assertion is partly correct for the modern markets that have emerged in Spain in recent decades where Norwegian exporters cannot operate fully with current Norwegian conventions. Initially, it was also claimed that the Norwegian producers' understanding of the Spanish market conventions was for the Norwegians priority to meet market requirements, or choose alternative marketing solutions. The discussion has shown that the latter most likely was the solution. Another part of the value chain therefore focuses on markets where Norwegian fish is a good commodity, especially as a raw material for processing traditional products for other markets. These traditional markets remain also in Spain, as well as in Portugal and Brazil; some Spanish firms refine the Norwegian product qualities for these markets. Consequently, the research assumption is correct in this respect. This discussion relies on the assumption that the Norwegian perceptions of Spanish market conventions and the Norwegian adaptation to own legal and industrial structure, link the Norwegians to traditional production and value chains.

In economic geography, local contexts exert a substantial influence as explanatory factors. In the geography of food, we assume local places create competitive advantages for producers through the impact of local culture on products. The findings of this survey show that local characteristics do not translate into product characteristics when the institutional framework of legal conventions on a more general level strictly regulates the value chain, thus preventing the emergence of local solutions for the benefit of generics. The Norwegian salted fish industry produces standard generic products with minimal differentiation to local market requirements. Traditional Norwegian production conventions will thus continue to dominate the Norwegian value chain, which bars the application of market conventions.

Although Norwegian producers have not been able to participate fully in the modern part of the salted fish value chain to Spain, discussion has shown that they are well aware of developments in the Spanish consumer markets. Seafood Norway has sought to develop this modern part of the value chain. This Norwegian promotion institution was in 2011 building on the Norwegian domestic conventions that will ensure the Spaniards their products without the use of phosphates, and consequently strengthen the HORECA branch of the value chain. However, such strategies do not compensate for continuous market presence and product adaptability.

References

Anfaco (Asociación Nacional de Fabricantes de Conservas de Pescados y Mariscos) (2010) Principles of sensory analysis in fish products. Presentation and demonstration of the results from the objective and sensorial test of the Norwegian salted cod. Report from the Spanish National Association of Fish and Seafood Canneries. Vigo, Spain, October 2010, unpublished manuscript

Benjaminsen H, Gabrielsen R (2010) Verdiskaping i form av videreforedlede produkter [Value adding in the form of semi-fabricated products]. In: Juliussen SS, Larsen HB, Lindkvist KB, Trondsen T (eds) Norsk Saltfisk i Spania [The Norwegian–Spanish Salted Fish Project]. Working paper.University of Bergen, Bergen

Boschma RA (2005) Proximity and innovation: a critical assessment. Reg Stud 39(1):61–74

Cornish SL (1997) Product innovation and the spatial dynamics of market intelligence: does proximity to markets matter? Econ Geogr 73(2):143–165

Crevoisier O (2004) The innovative milieus approach: toward a territorialized understanding of the economy? Econ Geogr 80(4):367–379

Dicken P (1998) Global shift: transforming the world economy. Paul Chapman, London

Drage Bakke MB (2010) Spanske konvensjonar og bygging av merkevare for norsk saltfisk [Spanish conventions and the building of trademarks for Norwegian salted fish]. Report 02/10. The Norwegian–Spanish Salted Fish Project. University of Bergen, Bergen

Espinosa A (2011) Regional market segmentation of a mature food product: the case study of salted codfish in Spain. The Norwegian–Spanish Salted Fish Project. Unpublished manuscript

Espinosa AS, Martinez IA (2010) Profiles of the Spanish salted fish markets. Report 02/10. The Norwegian-Spanish Salted Fish project, Bergen

Gereffi G, Humphrey J, Sturgeon T (2005) The governance of global value chains. Rev Int Polit Econ 12(1):78–104

Gertler MS (1997) The invention of regional culture. In: R. Lee, J. Wills (eds) Geographies of economics. Arnold, London, pp. 57–58

Gertler MS (2010) Rules of the game: the place of institutions in regional economic change. Reg Stud 44(1):1–15

Haugnes SE (2010) Consumers in industrial networks: A study of the Norwegian–Portuguese Bacalhau Network, Series of Dissertations 4/2010 BI. Norwegian School of Management

Lindkvist KB (2010) Mistrust and lack of market innovation: a case study of loss of competitiveness in a seafood industry. Eur Urb Reg Stud 17(1):31–43

Lindkvist KB, Røkenes A (2009) Med saltfisk på menyen. Fra en feltundersøkelse av saltfisk i restauranter i Barcelona [With salted fish on the menu. From a field survey of salted fish in restaurants in Barcelona]. Working paper 01/09. The Norwegian–Spanish Salted Fish Project. University of Bergen, Bergen.

Lindkvist KB, Sánchez JL (2008) Conventions and innovation: a comparison of two localized natural resource-based industries. Reg Stud 42(3):343–354

Lindkvist KB, Gallart-Jornet L, Stabell M, (2008) The restructuring of the Spanish salted fish market. Can Geogr/Le Géographe Canadien 52(1):105–120

Sánchez-Hernández JL (2011) The food value chain as a locus for (dis)agreement: Conventions and qualities in the Spanish wine and Norwegian salted-cod industries. Unpublished manuscript

Santana L (2010) Konvensjonspåvirkning og bygging av merkevarer for norske saltfiskprodukter i Spania. Vist gjennom eksempelstudiet West Fish [Convention influence and building brands for

Norwegian salted fish products in Spain. Presented through a case study of West Fish]. Report 03/10. The Norwegian–Spanish Salted Fish Project. University of Bergen, Bergen
Trondsen T (2012) Value chains, business conventions, and market adaptation: a comparative analysis of Norwegian and Icelandic fish exports. Can Geogr/Le G×eographe canadien 56(4):459–473
Xie J, Myrland Ø (2010) Modeling market structure of the Spanish salted fish market. Food Econ Acta Agric Scand, Sect C 7(2):119–127

Chapter 8
Innovation and Change of the Spanish Cod Fishing Industry

Manuel González-López, Alexandre Trigo and Sebastián Villasante

Abstract This chapter studies the changes which have occurred in the Spanish cod fishing industry in the last few years. Although innovation and change are usually understood to be key factors for economic success for both territories and industries, sometimes to innovate in a narrow sense is not enough. In some cases the survival of a given industry entails alteration of existing institutions and major routines. Although some authors have already analyzed this issue, more research is needed to understand in depth the relationship between innovation and institutions. Our results indicate that innovations implemented by the Spanish cod fishing industry go further than simple increments in its production system. The main driver of such deep changes is the restructuring which has happened in the Spanish salted fish market, the traditional market of the Spanish cod fishing industry. Social and cultural changes occurring in Spain during recent decades, together with an eruption of the strategies followed by strong foreign competitors, have changed the rules of the game in the market. As a result of restructuring of the Spanish salted fish market, the tiny Spanish cod fishing industry has been obliged to restructure itself in order to survive.

Keywords Spanish cod fishing industry · Production system · Innovation · Conventions · Institutional change

8.1 Introduction

The aim of this chapter is to analyze the present situation of the Spanish cod fishing industry as well as the changes it has experienced during the last few years. We also aim to understand the dynamics of industrial change and innovation in relation to institutions, particularly when referring to informal rules and conventions. Although innovation is usually understood to be a key factor for economic success, sometimes to innovate in a *narrow* sense is not enough. In some cases the survival of a given

M. González-López (✉) · A. Trigo · S. Villasante
Department of Applied Economics, Universidade de Santiago de Compostela ICEDE Research Group, Santiago de Compostela, Spain
e-mail: manuel.gonzalez.lopez@usc.es

K. B. Lindkvist, T. Trondsen (eds.), *Nordic-Iberian Cod Value Chains*,
MARE Publication Series 8, DOI 10.1007/978-3-319-16405-2_8

industry entails the alteration of existing institutions and major routines (David 1993). Although some authors have already analyzed this issue, more research is needed to understand in depth the relationship between innovation and institutions. Our study tries to add to such understanding as we analyze the case of small industry that has experienced deep changes affecting its entire production system, just in order to tackle with a new institutional, social and economic context.

The chapter is divided into the following parts: firstly, we briefly discuss the relationship between innovation and institutions from a theoretical viewpoint in order to achieve a better understanding of such a relationship. In the Sect. 8.2, a general description of the Spanish cod fishing industry is presented highlighting not only current features but also historical features. In Sect. 8.3, we briefly present the details of the research conducted in the Spanish cod fishing industry. In the following, we present the changes, which have occurred in the last few years in this industry, focusing on different types of innovations. We find that rather than incremental changes, the Spanish cod fishing industry has suffered a deep restructuring in the last decade affecting its entire production system. In Sect. 8.6 we try to bring about the ultimate reasons explaining those changes. Finally, some conclusions are included at the end of the paper that summarizes our main results and considerations.

8.2 Innovation, Industrial Change and Institutions

Industrial change is not only an economic matter but also a process where historical, technological, social and cultural factors play a central role. This evidence has been pointed out from different perspectives and academic fields. It is the case of the Evolutionary School (Nelson and Winter 1982) or a number of schools coming from the Economic Geography discipline like the New Industrial District stream (Becattini 1987), the Gremi School based on the concept of "milieux innovateurs" (Aydalot 1986; Crevoisier 2004), or the "flexible specialization" paradigm (Scott 1988; Piore and Sabel 1984). These contributions usually point out the systemic nature of economic processes where the actions of agents are shaped by conventions and institutions governing relationships within firms, sectors and society as a whole. The link between industries and territories is also a very important one since not a single economic behavior is universally found. Under this view, each territory and industry will show a specific pattern of development marked by historical, technological, social and cultural factors. Therefore, we need to look further than pure economic factors to understand economic processes.

Institutions, defined as "sets of commons habits, routines, established practices, rules, or laws that regulate the relations and interactions between individuals and groups" (Edquist and Johnson 1997, p. 46), play a central role in the process of economic change. Institutions, according to the referred authors, can be formal (such as laws and regulations) or informal (like conventions and habits). According to Carlsson and Jacobsson (1997), institutions stimulate innovation because they reduce uncertainties, coordinate the use of knowledge, mediate conflicts, and provide incentives. Nevertheless, such a relationship between institutions and innovation shows also an opposite side. Thus, to innovate also implies to alter existing routines, habits and even legal frameworks (Nelson and Winter 1982; Hayter 2004; Granovetter 2005), particularly in the case of radical innovations or important changes occurring in one industry. In this sense Douglas North (1994) has argued about the nature of economic change pointing out that whilst most of the time it is a matter of decisions based on existing routines and institutions, some changes might entail the alteration of such routines. These decisions: "*involve altering existing 'contracts' between individual and organizations. Sometimes that re-contracting can be accomplished within the existing structure of property rights and political rules, but sometimes new contracting forms require an alteration in the rules. Equally, norms of behavior that guide exchanges will gradually be modified or wither away. In both instances, institutions are being altered*" (North 1994, p. 361). Similar conclusions raised by David (1993), indicate that sometimes industries, in order to survive, must change their entire institutional framework upon which they are situated. This happens because in many circumstances the external environment pressure is so high that organizations and institutions collapse or dissolve.

When dealing with food industries like the one we analyze here we must take into account the deep changes that happened during the last decades in industrial, social and cultural institutions and conventions affecting this sector. Such changes have been largely discussed and analyzed by authors like Morgan et al (2006) referring to "worlds of food". In a similar line but referring to two specific food industries, Lindkvist and Sánchez (2008) have discussed the relationship between innovation conventions (non-formal institutions). The authors emphasize the importance of adaption to new market conventions and contexts for a firm survival. They conclude that those companies showing an adaptive behavior are more successful than those who show rigidity when changing their rules and internal conventions.

Therefore, we can summarize this section by pointing out that sometimes industries need to alter existing production systems in order to adapt to a new institutional context and to survive. This could mean a radical change in their production system or just partial changes affecting it. As we will try to show in the following paragraphs, the Spanish cod fishing industry has been involved during the last few years in a deep structural change. Such changes, which go beyond the simple (incremental) innovations in products or processes, have implied important changes in the entire production system and the rules governing it.

8.3 History and Present of the Spanish Cod Fishing Industry

The history of cod fishing in Spain is closely related to the grounds of Newfoundland and there are evidences about Basque vessels catching codfish in these grounds already in the sixteenth century. Nevertheless, cod fishing in modern times as a systematic activity begins during the first third of the Twentieth century. The industry was mainly located in two regions: Galicia and the Basque Country. In 1924, a

Galician trawler based in Vigo initiated its cod fishing activities in the waters of Newfoundland. Three years later, in 1927 the Basque company PYSBE, that had been established in 1919, started its fishing activities. In 1929, fishing in Icelandic and Northern European waters took place for the first time. The Spanish fleet made popular a particular fishing system consisting in "pair trawling" where two vessels where simultaneously used. As pointed out by Oya (1974) such was used for the first time in 1949 and it is still in use by some companies.[1]

On a global level, the Spanish fishing fleet became the third largest at the end of the 1960s with 80 trawlers. The catches reached 300,000 t at that time, mainly from the NAFO fisheries. As explained by Zeller and Pauly (2004), both stock declines due to over-exploitation and changes in the International Law of the Sea forced the reduction of captures of fleets like the Spanish one (Zeller and Pauly 2004). Since the establishment in 1977 of the Economic Exclusive Zones (EEZs) in the European Union (EU), Canada, United States and Norway, Spanish codfish quotes have progressively decreased; and at the end of that decade, catches had declined to less than 30,000 t.

Finally, the limitations established by the EU in the community fishing grounds meant in practice the closing of those grounds to the Spanish fleet. This happened just some years before the collapse of Northwest Atlantic Fisheries Organization's (NAFO) fisheries at the beginning of 1990s. Since then, the Spanish fleet has developed its activity in the Norwegian EEZ (based on the agreements between this country and the EU) and at the Svalbard fishing ground [2] (Fig. 8.1).

Despite its historical importance, the Spanish cod fishing industry is currently composed only of five companies that manage a fleet of nine active trawlers. At the beginning of 2009, there were six companies but one decided to stop its activities and transfer their fishing rights to another Spanish company. The remaining five companies share the current Spanish quota for codfish (around 12,000 t). Two of these companies locate their headquarters in Vigo (Galicia), while three are in San Sebastián (Basque Country). Nevertheless, some of the companies based in the Basque Country use Vigo as a base for their trawlers. Six vessels are based in the port of Vigo, while only three vessels are based in Pasaia Port, San Sebastián.

The economic importance of this fishing industry is quite small compared with the past or with the total Spanish fishing industry. Total employment of this industry has declined from 158 employees in 1998 to 79 employees in 2007. Total revenues, which are obviously determined by the codfish size quotes, have varied between € 20 and € 25 million in this period, while average revenues have varied between € 3.3 and € 4.5 million. A clear trend is not observed neither in total nor in individual revenues during the period. On the contrary, a clear increase in productivity (measured as revenue by employee) has taken place in the period (Table 8.1).

---

[1] Pair trawling is a fishing activity carried out by two boats, with one towing each warp. By using the towing power of two boats a larger net may be worked than would otherwise be possible, or alternatively, the two boats can share increased fuel efficiency.

[2] NAFO fisheries have recently re-opened although the quota given to the Spanish fleet is not large enough to initiate a fishing campaign there.

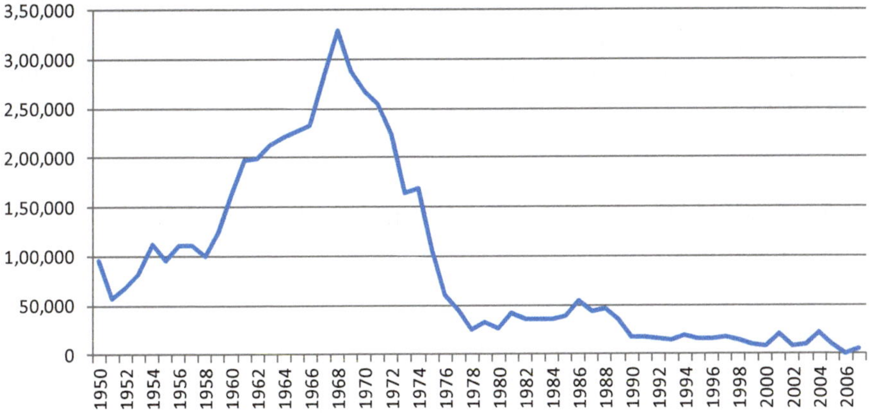

Fig. 8.1 Volume of production, codfish, Spain (tons). 1950–2006. (Source: Own elaboration based on FAOSTAT. Global production statistics 1950–2007)

Table 8.1 General information of the Spanish codfish trawler fleet

Company	Location of headquarters	Codfish quote share (%)	Number of vessels actively dedicated to cod fishing
Pesquera LaurakBat S.A.	*Pasaia—San Sebastián (Basque Country)*	*9.0*	*1*
Velaspex S.L.	*Pasaia—San Sebastián (Basque Country)*	*14.8*	*2*
Pesquera Rodríguez S.A. (Pescafria)	*San Sebastián (Basque Country)*	*27.4*	*2*
Pesquera Ancora S.L. (FormerTranspesca S.A.).	*Vigo (Galicia)*	*24.1*	*2*
Valiela S.A.	*Vigo (Galicia)*	*24.4*[a]	*2*

[a] Valiela has recently bought the quota share of León Marco S.A, the company that stopped its activities in 2009

8.4 Data Collection

We collected our data from a survey where a questionnaire was designed and interviews of relevant actors where carried out . As a first phase, the questionnaire was sent to the five companies of our sample but the response was quite poor since only one company returned the questionnaire (only partially filled out). Therefore, most of the information used in this report was obtained at the interviews where we used the same questionnaire as a structured guide. Four interviews were carried out at the companies' headquarters; three interviews took place in San Sebastián and one in Vigo. During the interviews, complemented with several phone calls, we captured relevant opinions and views from persons with a deep knowledge of the sector. We did not receive direct information from one of the companies (that refused to provide any information) nevertheless, as the sector is very small and all companies know each other, we gained information for the whole fleet and industry based on the other interviews.

8.5 Codfish Production: Traditional Systems and Recent Changes

The Traditional Method of Salting Fish and its Associated Production System The Spanish codfish industry developed a particular method of conservation based on salting the fish on-board the vessels. The process was developed at the same time as long distance cod fishing, starting in Spain at the beginning of the sixteenth century according to historians. This method for salting cod, also known as dry salting or kench curing, consists in rubbing the solid salt into the fish meat which is then pile-stacked with alternate layers of fish and salt. Due to increasing pressure from the overburden at the bottom of the cod stacks, this method results in the dehydration of much of the salted fish. Lindkvist et al. (2008) have indicated that this pile salting method arose to optimize the vessels' stock capacity since the space on board schooners is limited. This aspect is particular relevant when dealing with long-distance fishing. Once the fishing activity ended, the salted fish was landed mainly in Galician ports with adequate infrastructure and trained labour. The main port for codfish landing was (and still is) located in the town of Cangas, near Vigo, where it was temporarily stocked before sending it to driers situated in other parts of Spain with an appropriately dry climate.

The conservation and technological system described above was also closely related to specific market relationships and conventions. In fact, we can affirm that they form all together a particular production system. Thus, the Spanish fishing industry commercialized its "bacalao" to dealers and processers that acted as intermediaries in the Spanish retail market. Consumers could buy the salted fish in local markets, small shops, supermarkets and restaurants that were served by those intermediaries. In many cases there were long-term relationships based on mutual trusts between the fishing industry and the clients in such a way that product was already sold in advance (previously to be fished). Finally, the product, shaped by the specificities of the production method, was very popular in the Spanish and Portuguese market. It showed a particular yellow-green colour due to the oxidation process, appearance (split or butterfly) and taste. A specific preparation procedure was also required for cooking consisting in desalting and dehydrating the salted fish a couple of days before cooking. As pointed out by Lindkvist et al. (2008), these conventions were so established that even after the collapse of the Spanish fishing industry, foreign producers who exported salted fish to the Spanish market followed them.

Recent Changes in the Production System of the Spanish Trawler Fleet The traditional salted fish production system has been progressively abandoned. Now three companies still maintain this system even though they combine it with freezing. Overall, in 2009 less than a third of the catches of the Spanish fleet used the traditional on-board system. Nevertheless, despite substitution of the traditional salting method by freezing, a significant share of the codfish is still sold as salted to the final consumer, since it is thawed and salted by the processers. This happens particularly with catches landed in Norway where at least two companies land part of their catches. Besides the changes regarding conservation, the Spanish companies have carried out product innovations. Thus, they are currently selling three main types of products: (a) Split cod or "Butterfly". The basic, traditional product, sold as salted fish but in some cases landed as frozen cod and salted on land. (b) HG (headed and gutted) sold mainly as frozen but where the processors as in the previous case also are thawing and salting some parts. (c) Fillets sold by the fishing firms mostly as frozen. Together with these basic products at least one company is producing little cod loins, "kokotxas" (cod cheeks), and other products. Besides, almost all companies have implemented minor changes in product appearance in order to attend to market requirements.

These innovation affecting products and conservation processes have also required innovations in other spheres. Thus, one company has built a completely new trawler and another has made a deep transformation in their vessels. These transformations were particularly relevant at the on-board processing plants and refer to new industrial cold systems, new packing methods and, in general the design of more efficient plants. It is important to note, as pointed out by the interviewees, that most technologies incorporated were already available and commonly used in vessels dedicated to fish and produce other species. Therefore we are not dealing here with limitations from the technological viewpoint but instead to market factors (eventually overcome) that made it unnecessary to incorporate those technologies into the Spanish codfish industry.

Market Changes As we have mentioned, the Spanish industry traditionally sold their "bacalao" to dealers or industrial processers that at the same time sell the product to retailers. Nevertheless, important changes have recently happened also in this regard and new commercializing channels have emerged progressively substituting the old ones. Firstly, some companies have diversified their clients, moving apart from dealers and processers to retail chains, not only Spanish but from other countries like France or the United Kingdom. This applies particularly for codfish commercialized fillets, which in the UK are sold under the famous dish of "fish and chips".

The previous change led to a second relevant change since an internationalization process of sales has been going on. As mentioned, some companies (at least two) are now selling to clients situated in countries other than Spain and the largest company of the industry) actually sells the majority of its captures abroad. Ultimately, in 2009 around 41 % took place at international markets, according to the information obtained from the companies. An important part of foreign sales take place at the Norwegian ports since the cod landed in Norway is usually a frozen cod product that is later salted (and in many cases exported to Spain).

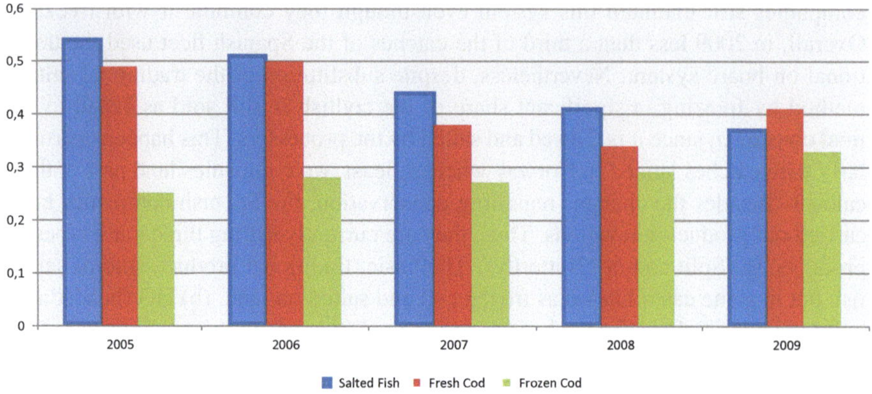

Fig. 8.2 Codfish consumption per capita (Kgs/population). (Source: Own elaboration based on the Base de datos de consumo en hogares of Ministerio de Medio Ambiente y Medio Rural y Marino)

Therefore, we can conclude that changes in the Spanish codfish industry have also affected the market sphere so the whole production system has changed. Whilst the traditional salted fish was sold exclusively to Spanish intermediaries, the new products are sold mainly to foreign clients, some of them international retail branches.

8.6 What Has Made the Spanish Cod Fishing Industry Change?

We have seen how the Spanish codfish industry has introduced deep changes regarding both production and commercialization. The result is the progressive abandoning of the traditional on-board salted fish, the production system that has characterized this industry since the beginning of the last century. The reasons are closely related to the strong restructuring occurring in the Spanish salted fish market over recent decades, a process that has been described in depth by Lindkvist et al. (2008). Firstly, there is a general trend towards the introduction of frozen products into consumer diets due to social and cultural changes. This trend is mainly due to the incorporation of women to the labor market, which has reduced the time for home tasks usually conducted by women in traditional societies and therefore has made frozen products become very popular.[3] This trend applies also for the case of codfish. In Fig. 8.2, we can observe how frozen codfish is increasing its quota in the Spanish market while salted fish (where salted codfish is included together with other minor species) has reduced it. Consumption of codfish in a frozen basis has also meant changes in its presentation and taste, different from the traditional salted one. Of course, there still exists a market for traditional salted and, as pointed by Espinosa and Bjønes in this

[3] According to official labor statistics, the participation rate of women in the labor market moved from 275 % in 1977 to 45 % in 2004 (INE 2013).

volume, a regional segmentation of the market exists. Nevertheless, the underlying tendency for all of Spain towards frozen seafood products, in similarity to many other countries, seems to be quite clear.[4]

Moreover, the traditional on-board salted fish shows some other comparative disadvantages in this new scenario. For instance, salted fish produced by a means of alternative salting methods, like direct brining, is now dominant in the Spanish market. These methods, largely introduced by foreign producers (mainly Icelandic), are more flexible for both production and consumption. The desalting and dehydrating process are unnecessary for consumers (since the producers do it). These new products have continuously grown in the Spanish market and somehow they have become standards in many senses. An example is the appearance, and particularly the white colour, as it is associated with freshness and quality in the case of is fish products. In this line, light-salted fillets shown a much "friendly" appearance for the consumer since new production method conserve (or cause) the white colour while on-board salting usually gives codfish a yellow appearance (this is the main reason why in some places it was known as golden codfish). Although the Spanish producers insist in the higher quality of their product and maintain that the traditional salted fish does not show cost disadvantages, they are aware, as pointed by one of the interviewees, that "*it is too late to launch any strategy to maintain the traditional salted fish*".

Finally, social, cultural and economic changes have also altered the traditional distribution channels of the Spanish codfish. Large supermarkets and retail chains have become the key stakeholder in the Spanish food market displacing the traditional dealers and processers that were dominant in the market of salted fish. According to Lindkvist et al. (2008), this process has also been influenced by the Icelandic producers' strategies that, established their own sales network, avoiding the traditional Spanish one. According to these authors, introduction of salted frozen fillets by Icelandic producers explains the appearance of new distributions systems created to meet new demands in the Spanish market.

8.7 Conclusions

In this chapter, we have analyzed the present situation of the Spanish cod fishing industry as well as the changes that have been experienced during the last few years. In order to study these changes we conducted a research to investigate the innovative patterns followed by this industry in the last decade. Innovations are understood here not only from a technological perspective but also from a rather systemic viewpoint, i.e. as the response of a particular industry to general changes affecting social or cultural factors. Our results indicate that innovations implemented by the Spanish cod fishing industry go further than simple incremental innovations in products or

[4] We know that part of frozen codfish has previously been subject to a salting process, as is the case of the light salted frozen fillets. In any case it seems clear that frozen codfish is increasingly present in the Spanish market regardless of whether it has been salted or not.

processes but are accompanied by a deep change in the entire production system. The main driver of such deep changes is the competitive pressure from the restructuring, which has happened in the Spanish salted fish market, the traditional market of the Spanish cod fishing industry. At the same time, we have seen how such restructuring is closely related to social and cultural changes both on the consumption side and in access to fishing grounds occurring in Spain during recent decades. The eruption of the strategies followed by strong foreign competitors with strong access to fishing quotas is another major explaining factor, since they also have been able to change the rules of the game in the market (Lindkvist et al. 2008).

Because of the restructuring of the Spanish salted fish market, the tiny Spanish cod fishing industry has been obliged to restructure itself in order to survive, constrained by the value chain structure starting with a distance trawler fleet. This has been done by different means encompassing process, product and market innovations. Firstly, they have been forced to incorporate the freezing conservation system and progressively abandoning the on-board salting system (process innovation). Secondly, they have diversified its product offer by incorporating new products like salted/frozen fillets and others (product innovation) possible to process from the onboard frozen catches. Finally, they have modified their market strategies both internationalizing their sales and selling directly to retailer and supermarket chains.

References

Aydalot P (1986) Milieux innovateurs en Europe. GREMI, Paris

Becattini G (1987) Mercato e forze locali: il distretto industrial. Il Mulino, Bologna

Carlsson B, Jacobsson S (1997) Diversity creation and technological systems: a technology policy perspective. In: Edquist C (eds) Systems of innovation: technologies, institutions, and organizations. Pinter, London, pp 266–294

Crevoisier O (2004). The innovative milieus approach: toward a territorialized understanding of the economy? Econ Geogr 80(4):367–379

David PA (1993) Why are institutions the 'carriers of history'?: path dependence and the evolution of conventions, organizations and institutions. Struct Chang Econ Dyn 5(2):205–220

Edquist C, Johnson B (1997) Institutions and organizations in systems of innovation. In: Edquist C (eds) Systems of innovation, technologies, institutions and organizations. Pinter, Oxon, pp 41–63

FAOSTAT (2009) Global production statistics 1950–2007. http://www.fao.org/fishery/statistics/global-production/en

Granovetter M (2005) The impact of social structure on economic outcomes. J Econ Outcomes 19(1):35–50

Hayter R (2004) Economic geography as dissenting institutionalism: the embeddedness, evolution and differentiation of regions. Geogr Ann 86(B):95–115

INE (2013) Encuesta sobre la población activa. Varios años. Instituto Nacional de Estadística

Lindkvist KB, Sánchez JL (2008) Conventions and innovation: a comparison of two localized natural resource-based industries. Reg Stud 42(3):343–354

Lindkvist B, Gallart-Jornet L, Stabell MC (2008) The restructuring of the Spanish salted fish market. Can Geogr 52(1):105–120

Ministeriode Medio Ambiente y Medio Marino (2009) Bases de datos de consumo de hogares. http://www.magrama.gob.es/es/alimentacion/temas/consumo-y-comercializacion-y-distribucion-alimentaria/panel-de-consumo-alimentario/base-de-datos-de-consumo-en-hogares/consulta.asp
Morgan K, Marsden T, Murdoch J (2006) Worlds of food: place, power, and provenance in the food chain. Oxford University Press, Oxford
Nelson R, Winter S (1982) An evolutionary theory of economic change. Harvard University Press, Cambridge
North D (1994) Economic performance through time. Am Econ Rev 84(3):359–368
Oya J (1974) Las pesquerías españolas en onda larga: la "gran pesca" del bacalao en los bancos del Atlántico noroccidental. Rev Geogr 10:48–78
Piore MJ, Sabel CF (1984) The second industrial divide: possibilities for prosperity. Basic Books, New York
Scott AJ (1988) New industrial spaces: flexible production organization and regional development in North America and Western Europe. Pion, London
Zeller D, Pauly D (2004) The future of fisheries: from 'exclusive' resource policy to 'inclusive' public policy: perspectives on eco-system-based approaches to the management of marine resources. Mar Ecol Prog ser 274:295–303

Chapter 9
Value Chain Policy, Industrial Conventions and Market Performance: A Comparative Analysis of Norwegian and Icelandic Cod Exports to Spain

Torbjørn Trondsen

Abstract This paper identifies and explains the differences in market performance of value chains for Norwegian and Icelandic Atlantic cod products exported to Spain from 1990 to 2008. Iceland had a higher value performance than Norway in the chain linking the coastal fishing fleet to fresh fish auctions and in the vertically linked chain from trawler fleet to processing industries. Norway, however, has achieved higher product value in the industrial fishing fleet that can sell their catches through a law-protected value chain system that includes minimum prices and frozen fish auctions. These performance differences are shown associated with the two countries' value chain policy and regulations that have been shaping the direction of industrial conventions over time.

Keywords Export trends and Norwegian and Icelandic shares · Value adding performance · Norwegian and Icelandic fish processing industries · Differences in value chain structures and conventions

9.1 Introduction

Value chains stress the interdependencies of economic activities across space and raise questions about location choice, organizational structures, markets and the role of policies and regulations, and they are widely cited in economic geography, industrial economics and marketing (e.g. Porter 1990; Dicken 1998; Wathne and Heide 2004). This paper contributes to this literature by a comparative analysis of two closely related value chains, namely the Norwegian and Icelandic codfish value chains that serve Spanish markets. This type of comparison is especially useful

T. Trondsen (✉)
Norwegian College of Fishery Science, University of Tromsø, Tromsø, Norway
Tel.: +4777645567
e-mail: torbjorn.trondsen@uit.no

Department of Geography, University of Bergen, Bergen, Norway

K. B. Lindkvist, T. Trondsen (eds.), *Nordic-Iberian Cod Value Chains*,
MARE Publication Series 8, DOI 10.1007/978-3-319-16405-2_9

for revealing the implications of policy, and the role of location and organizational choices.

Most businesses operate in value chain structures and market networks that bring products and services from the stage of input raw materials to the final market according to demand. The seafood value chains provide a good empirical background for studying the role of markets and the process of business orientation in a structure where strongly regulated natural resources are turned into products ready for the consumer's dinner table. Even for a common raw material and closely related consumer products, however, the institutions and conventions that bind value chains differ. These differences underline that the nature of value chains involves questions of choice and policy that in turn have implications for business performance and local development.

This paper analyses the differences between the development paths of the two codfish value chains. Earlier studies have shown significant differences between Norway and Iceland in the realization of market values of exports to Spain of similar salted cod products in terms of both unit values and growth (Trondsen 1995; Lindkvist et al. 2008). The research questions in this paper are: What are the structural and institutional differences between the two cod value chains? Are there differences in export performance between the chains? Do the differences reflect differences in business conventions and in the two societies' regulation of fisheries management conventions?

Conceptually, this analysis uses a Structure–Convention–Performance (SCP) value chain model (Chap. 2) that modifies the classic SCP model of industrial organization economics by replacing "conduct" with Salais and Storper's (1992) business "convention". Salais and Storper define a "convention" as an accepted way of thinking, practices, agreements and their associated informal and institutional forms that are bound together by mutual expectations. This definition is an extension of the concept of "conduct" in the SCP model, which is limited to the actual organizing and doing of something. The conventional interpretation of structure is retained, based on forms of industrial organization and degrees of competition, while performance is ultimately interpreted in terms of market share and value in final consumer markets. This paper has two main parts. The first part presents the methodology and empirical findings. The second part characterizes and discusses the similarities and differences between the Icelandic and Norwegian value chain performances and conventions.

9.2 Methodology

The main scope of the study is the actual flows of products and money throughout the value chain between markets for cod products and supply from wild cod harvesting landed in Iceland and Norway, facilitated by the institutional structures embedding in the value chains in the two countries. The analysed data are collected by interviews with industry actors, direct observations of production chains and official comparable trade data of cod landings and exports from Iceland and Norway.

Export volumes and average export values are proxies for export performance of the cod products in question. All landings, import and export data are converted into live weight equivalents in order to compare the export value of different cod products. The main export products with different conversion rates from live fish are whole gutted fish head-on (as in Iceland) or head-off (as in Norway) chilled on ice, round frozen or dried (stockfish); fillets that are fresh chilled, salted, frozen or light salted frozen and split fish that are wet salted or dried salted (clipfish). All products must, before consumption, be further processed to consumer-ready products by cutting and packaging fresh, frozen and salted products or dewatered salted fish and clipfish and thawed frozen fish. All these different product forms have their own specialized value chains.

Statistics Norway (SSB) and Statistics Iceland (Hagstofa Islands) have provided primary landing data and Seafood Norway (EFF) has provided Norway Statistics (SSB) data for export and import in the European Union (EU), Norway and Iceland. The calculation of the live weight equivalent from product weight follows the Norwegian Directorate of Fisheries' official conversion factors .

All values are converted into 2008 Euro values according to the official EU conversion rate in order to compare average values paid for cod at different stages in the value chain from landings and export.

The qualitative description of management conventions in the two countries are based on scientific literature, reports, interviews with key Norwegian and Icelandic informants (Trondsen and Arnarson 2009), supported by the author's observations of the cod value chain over a 40 year period from positions as an active production manager, student, researcher and industry consultant.

9.2.1 Norwegian and Icelandic Cod Exports to Spain

The Norwegian and Icelandic exports of cod to Spain, measured in volume and average price per kilo live weight, show a significant Icelandic expansion in recent decades (Table 9.1). Thus, the volume of Icelandic cod exports to the Spanish market from the 1990s to the 2000s increased by 42 % while the total Icelandic global cod exports decreased by 11 %. As a result, Spain increased its share of total Icelandic cod export volume from 14 % in the 1990s to 22 % in the 2000s. Norwegian exports to Spain have, in the same period, decreased by 21 %. The share of Norway's cod exports to Spain, however, remained the same (5 %) as Norway's export volumes declined in general. Iceland has therefore strengthened its market position in the Spanish cod market.

Iceland received a higher average value of cod products exported to Spain compared with Norway in both the 1990s and the 2000s (Table 9.1). However, the difference has been declining. The average value of Norwegian exports to Spain in the 1990s was only 92 % of the comparable Icelandic average value but increased to 98 % in the 2000s, which is the same percentage of the value per kilo Norway obtained of its total cod export compared to the total Iceland export. The average

Table 9.1 Norwegian and Icelandic cod exports to Spain and globally 1990–2008

Annual cod exports to Spain 1990–2008. Average live weight in metric tons and unit value in 2008 €

	Norway		Iceland		Norway in % of Iceland	
	1990–1999	2000–2008	1990–1999	2000–2008	1990–1999	2000–2008
Cod exports to Spain	23,327	18,377	38,294	54,238	61	34
Change 90s–00s		(− 21 %)		+ 42 %		
Total cod export	453,477	345,437	281,745	250,970	161	138
Change 90s–00s		(− 24 %)		(− 11 %)		
Export share of Spain	5 %	5 %	14 %	22 %		
Export value/kg to Spain	1.74	2.00	1.89	2.03	92	98
Change 90s–00s		+ 15 %		+ 7 %		
Total export value/kg	1.71	2.12	1.74	2.17	98	98
Change 90s–00s		+ 24 %		25 %		

value of Norwegian exports to Spain in the 2000s increased by 15 % compared to the 1990s, which is lower than the 24 % increase in the average value of total Norwegian cod exports. Iceland's average total value of exports of cod products increased at the same rate as for Norway. Lower priced products have thus driven the Icelandic export expansion in the Spanish market.

9.2.2 *Export Trends and Shares*

The total cod landings are driving the export volumes (Fig. 9.1). The larger annual variation in Norwegian landings make the main difference between Norway's and Iceland's export profiles, which developed strongly from about 200,000 MT (metric tons) in 1990 to 620,000 MT in 1997 and down again to 300,000 MT in 2008, while Icelandic annual landings varied between 200,000 and 260,000 MT. The Norwegian variations in landings were driven by growing cod stocks in the 1990s and, from the early 1990s–1998, and further driven by increasing Russian landings after the Soviet Union was transformed into Russia, as shown in Fig. 9.2 (see also Table 9.3 below).

Figure 9.3 shows the product mix of Icelandic cod exported to Spain. The main export growth was steady until it peaked in 2005. Frozen fillets, mainly light salted,

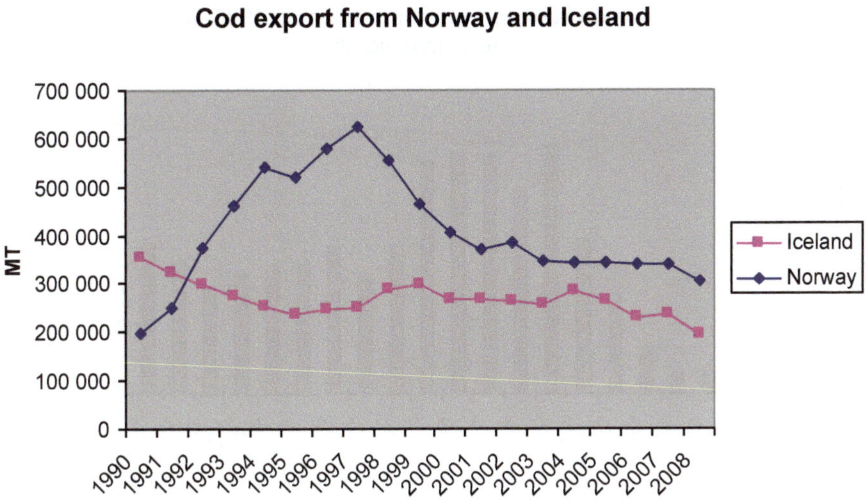

Fig. 9.1 Total cod exports from Norway and Iceland 1990–2008 in metric tonnage live weight

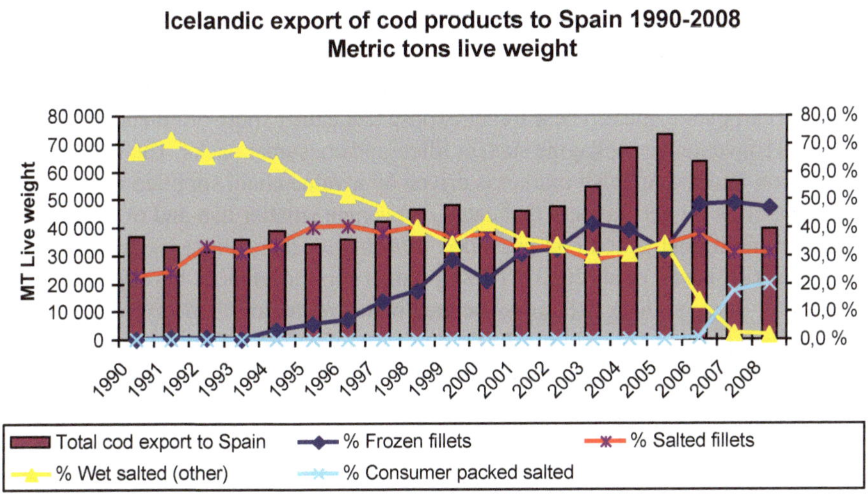

Fig. 9.2 Total Icelandic exports 1990–2008 to Spain divided into product categories

was the main export product which has grown to an export share of about 50 %, while traditional wet salted cod's share of exports has fallen steeply from 70 % in the early 1990s to 30 % in 2005. After 2005, a new shift occurred in which consumer packed salted cod increased its export share to almost 20 % at the expense of the traditional wet salted cod, which, in the export statistics, was almost zero in 2008. Iceland's export share to Spain of salted cod fillets has been stable at about 30–40 % of the total quantity exported. The recovery rate of salted fillet from the live weight is lower than for wet salted fish, which includes the bone and belly parts of the fish. Iceland

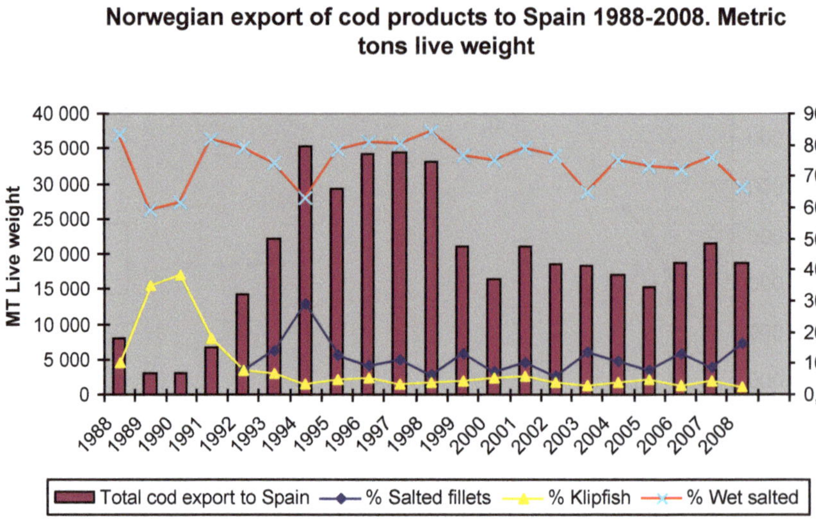

Fig. 9.3 Total Norwegian exports to Spain 1988–2008 divided into product categories

has compensated for this reduced recovery rate by producing special products such as salted belly and dried backbones (for the Nigerian market) cut from the fillets but that still carry some meat. These trends expose Icelandic export-oriented conventions towards convenience packaging such as fillets and consumer packs. The frozen fillets expansion into the Spanish market is driven by a value chain supplied upstream by fish caught by bigger trawlers that catch on average smaller fish and of lower value than the traditional salted cod produced from bigger fish caught by selective long line gear (Trondsen et al. 2003). These product mix differences between the value chains for thin and thick fish are associated with a reduction in the average value of Icelandic exports to Spain in the 2000s compared with the 1990s.

Figure 9.3 shows a similar Norwegian profile of exports to Spain, where the exported quantity represents a stable share of the total Norwegian cod export as shown in Table 9.1. The product mix exported to Spain is stable: wet salted cod represents between 65–80 %, while the export shares of fillets are 8–12 %, except in 1994 when the fillet share was 30 %. Norwegian exporters are not, to the same degree as Icelandic exporters, involved in the value chains for salted belly and backbone as by-products from fillet production. These by-products go mainly into lower value animal feed value chains. Iceland therefore makes higher value added products from the limited live cod-fish quotas by engaging in more specialized value chains compared with their Norwegian competitors. The Spanish market represents, however, only 5 % of the Norwegian and 22 % of the Icelandic total global cod quantities exported. Exporters, therefore, have alternative export value chains to play with.

Table 9.2 shows that the Norwegian cod products export profile is more focused on salt fish value chains than is the Icelandic. Iceland exports 55 % and Norway 28 %

Table 9.2 Norwegian and Icelandic export shares and average export values (2008 € value) of cod products in live weight ranked by 2000–08 average value

Average export share (ExS) and export value (€ /kg) of all cod products in live weight in 2008 €

Norway to	1990–1999		2000–2008		Iceland to	1990–1999		2000–2008	
	ExS (%)	€ /kg	ExS (%)	€ /kg		ExS (%)	€ /kg	ExS (%)	€ /kg
Italy	11	2.1	11	2.3	France	11	1.7	5	2.6
France	7	1.6	7	2.3	Italy	5	2.1	3	2.5
Denmark	7	1.5	7	2.2	Portugal	9	1.8	13	2.5
Brazil	6	2.2	6	2.2	Other markets	57	1.7	55	2.2
Spain	5	1.8	5	2.1	Greece	2	1.8	2	2.1
Other markets	33	1.7	28	2,0	Spain	14	1.9	22	2,0
Portugal	29	1.8	33	2,0	Brazil	< 0.5		< 0.5	
Greece	2	1.7	2	1.9	Denmark	2	1.6	< 0.5	
All markets	100	1.7	100	2.1	All markets	100	1.7	100	2.2

of their cod to other markets such as the United States and the United Kingdom, rather than the main salt fish markets. Both countries have strengthened their focus on salted fish markets from the 1990s to the 2000s shown in the reduction of their export shares in other markets from 33 to 28 % in the Norwegian case and from 57 to 55 % in the Icelandic case. Norway has, compared with Iceland, a bigger export share in Italy, Portugal and Brazil. The Norwegian–Italian value chain relies to a high degree on stockfish markets where especially Lofoten skrei (spawning cod) has a unique position. The Norwegian–Portuguese and Norwegian–Brazilian value chains rely mainly on Norwegian clipfish industry's historic position in these markets. Iceland has not entered the clipfish industry and has left these markets to the Norwegians. However, the Icelandic export share to Portugal, the biggest clipfish market, increased from 9 to 13 % on average from the 1990s to the 2000s, partly related to the build-up of a clipfish industry in Portugal demanding wet salted fish as raw material. However, an interesting observation is the fact that Iceland at the same time increased both their export share and their average export value to Portugal from € 1.8 to € 2.5/kg while Norway increased its export value only from € 1.8 to € 2.0/kg. This price difference may indicate that Iceland is exporting more of the more highly valued thick wet salted fish caught by long-line fishing gear to Portugal while Norway is exporting all types of clipfish qualities and sizes which, on average, command lower prices.

Table 9.2 also shows that Iceland, compared with Norway in the 2000s, has gained a price advantage per kilo in France (€ 2.6 against € 2.3) in Portugal (€ 2.5 against € 2.0) and Italy (€ 2.5 against € 2.3). This has in sum contributed to the total Icelandic average export values for cod being 2 % higher than Norwegian export values in both these periods (Table 9.1).

9.2.3 *Value Adding Performance in the Value Chains*

Table 9.3 shows the product flow and trading margins in the Norwegian and Icelandic value chains. One main difference between Norway and Iceland is the size of the Norwegian source raw material caused by the huge import of cod from Russian vessels' direct landings, accounting for one-third of the Norwegian exported cod. The Russian landing pattern for cod has changed from fresh in the 1990s and frozen in the 2000s. The figures show that, while the Norwegian own-landings of cod only was 14 % and 7 % higher than Iceland in the 1990s and the 2000s, respectively, the total supply of cod, including Russian landings in Norway, available for export was on average 62 % higher in the 1990s and 49 % higher in the 2000s compared with Iceland.

The figures also show that the landing value per kilo was on average 11 % and 7 % higher in Norway compared with Iceland in the 1990s and 2000s, respectively, while the export value per kilo was 2 % lower in Norway in both periods. A calculation of the export value per kilo landings also shows a 2 % higher Icelandic value in the 1990s but 10 % higher in the 2000s, which indicates that the Icelandic industry has been

Table 9.3 Supply flow, and processing and exporting margins in the Norwegian and Icelandic supply chain for cod 1990–2008

Cod total supply, domestic consumption and exports. Live weight in metric tons and unit value in 2008 €						
	Norway		Iceland		Norway/Iceland	
	Average 1990–1999	Average 2000–2008	Average 1990–1999	Average 2000–2008	Average 1990–1999	Average 2000–2008
Cod landings	285,995	220,493	250,044	207,014	114 %	107 %
Cod imports	137,265	91,943	11,676	2,846	1,176 %	3,231 %
Total cod supply	423,260	312,436	261,720	209,860	162 %	149 %
Import of landings	32.4 %	29.4 %	4.4 %	1.3 %		
Supply value/kg	1.39	1.85	1.25	1.73	111 %	107 %
Total cod exports	453,477	345,437	281,745	250,970	161 %	138 %
Export of landings	109.5 %	110.6 %	107.6 %	119.5 %		
Import of exports	30.3 %	26.6 %	4.4 %	1.1 %		
Export value/kg export	1.71	2.12	1.74	2.17	98 %	98 %
Export value/kg landing	1.83	2.34	1.87	2.59	98 %	90 %
Trading margin	23 %	15 %	39 %	25 %		

Table 9.4 The Norwegian and Icelandic fish processing industries' cost of raw fish, salary and contribution margin in percentage of sales value 1997–2008. The difference between the sum of the three categories and all costs is calculated to other operational costs. (Data: Bendiksen 2004, 2010; Statistics Iceland 2010)

	1997–2003		2004–2008	
	Norway	Iceland	Norway	Iceland
All				
Raw fish	73	59	74	56
Salaries	12	15	11	16
Operational margin	4	11	4	11
Freezing				
Raw fish	76	51	76	52
Salaries	11	22	12	21
Operational margin	3	12	3	12
Salting or drying				
Raw fish	80	66	79	58
Salaries	8	13	9	15
Operational margin	4	8	4	12

increasing the value margin it generates from its limited codfish resources compared with Norway. These calculations also show that the Icelandic cod processing and exporting industry on average had 16 % (1990s) and 11 % (2000s) higher trading margins compared with their Norwegian competitors.

Table 9.4, based on industrial data; confirms the findings in Table 9.3 based on trade data, comparing the allocation of operational costs and contribution margin in the Norwegian and the Icelandic fish processing industries. The table shows that the Icelandic industry on average has an operational margin (which is financing the investment costs and profit) of 11 % of sales value compared with 4 % in the Norwegian fish processing industry during 1997–2008. Similar figures for freezing processors are 12 % in Iceland and 3 % in Norway and for salting and drying processing 8–12 % in Iceland and 4 % in Norway. Table 9.4 also shows the reason for this difference: The raw material costs of sales value in Iceland are only 56–59 % against the Norwegian 73–74 %, but the salary costs are higher; 15–16 % of the sales value against the Norwegian 11–12 %. The cost differences are even higher for freezing, where the Icelandic processing plants on average pay 51–52 % of the sales value while the Norwegian processing industry pays 76 %. The differences are also high in the processing of salt and drying fish, where the raw material costs in Norwegian plants are 79–80 % and salaries 8–9 % of sales value while the Icelandic processors pay 58–66 % in raw material costs and 13–15 % in salaries of sales value.

The large Russian landing share of the total Norwegian export (30 % in the 1990s and 27 % in the 2000s) has however put pressure on average export prices and trading margins, which partly may explain the differences. The Russian landings frozen on board the vessels are imported into freezing storage and partly exported directly, headed and gutted (H&G) at very low trading margins. The EU import duties for Russian fish are between 7.5 % for frozen cod fillets and up to 20 % for salted cod fillets, while the Norwegian cod take advantage of the EEA (European Economic Area) agreement that allows for duty free cod exports to the EU.

Table 9.3 also shows a large discrepancy between landings and exports figures calculated in live weight equivalents. Norway's exports are about 10 % higher than the recorded landed cod quantity, while Iceland's was 8 % higher in the 1990s and 19.5 % higher in the 2000s. The official Norwegian conversion factors from product to live weight are applied. Either some landings are not registered or the recovery of product weight relative to live weight is much higher than the official list (see discussion later).

9.2.4 Differences in Value Chain Structures and Conventions

The main findings from the presented figures show that Iceland's cod value chains are doing better than Norway's in market-oriented value adding performance, which, according to Fig. 2.1, Chap. 2, may relate to differences in the countries governance and regulatory convention structures.

Figure 9.4 shows the differences between the Icelandic and Norwegian value chain structures. Iceland exports a diversified portfolio of product qualities according to preferences and value adding opportunities in each market segment. This structure has on the supply side, been driven by the fresh fish auction for multi-species landings. The auction is serving both the coastal fleet and the vertical integrated companies controlling offshore vessels and processing plants. The latter are mainly supplied by their own fleet, but they are improving their hieratic scale specialization by sending to auction surplus species and fish sizes in their landings that do not fit into their specialized production lines and then purchasing from the auction species and fish sizes that fit into their production lines; such transactions improve production efficiency.

Live fish catches are limited and subject to quotas; maximizing values from the cod landings requires an optimal value adding to all parts of the fish. The Icelandic auction system makes flexible specialization conventions possible in the production so that the product–market mix can be optimized by sorting, standardizing and packaging of products according to sizes, species, product forms and conservation forms as fresh, salted, frozen or light salted frozen according to market segment requirements (Trondsen and Arnarson 2009). To make profit in the competitive pressure the fish auction represents, the processing industry must add value to all parts of the fish by use of a broad portfolio of many alternative markets and product forms.

(Money flow -----------, Physical flow ———)

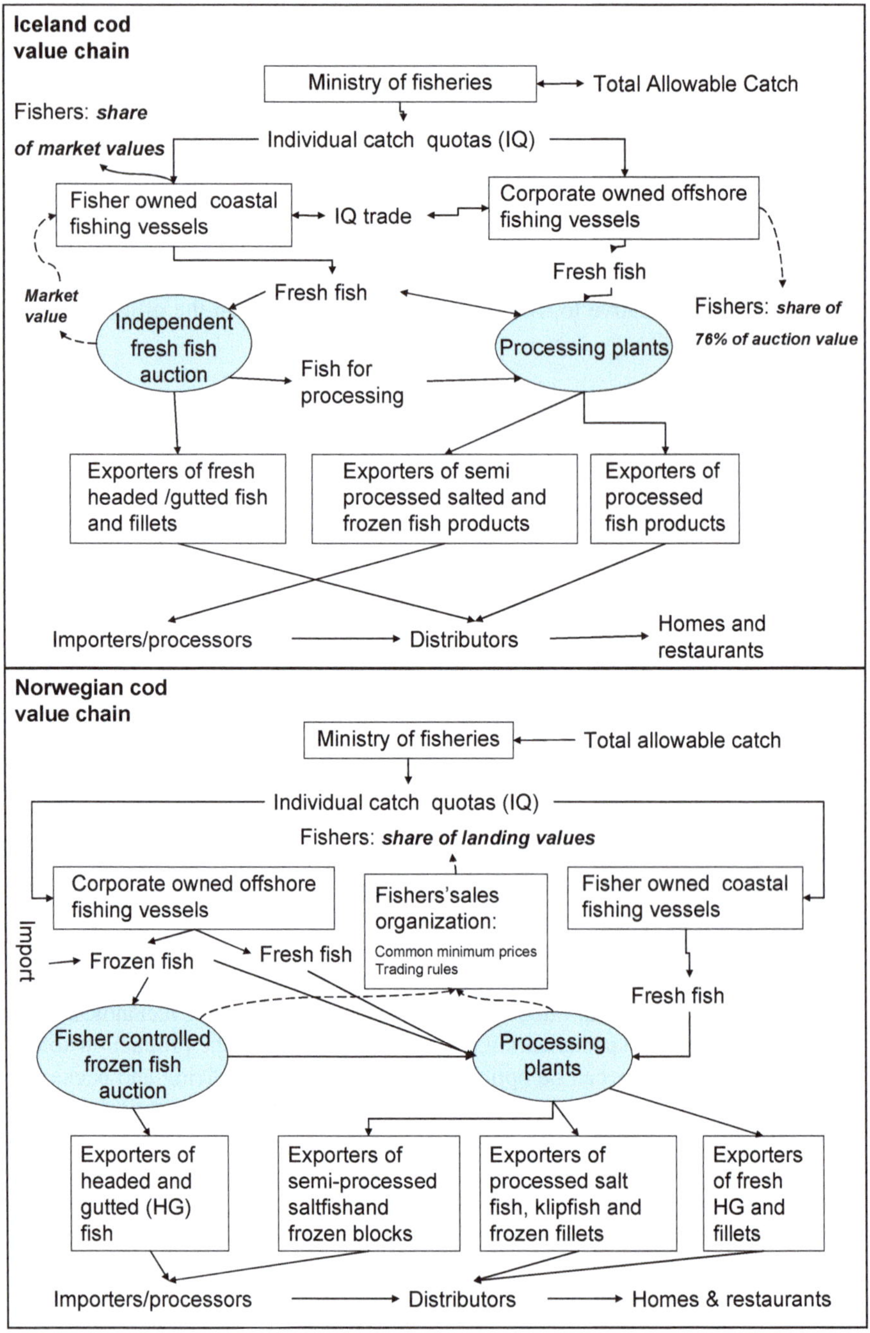

Fig. 9.4 Icelandic and Norwegian value chains for cod and cod products. (Money flow ———, Physical flow ———)

Prices in Iceland are set in the fresh fish auction where buyers are free to purchase only the species or quality they want. On average, only 17 % of Icelandic landings are sold at auction, but these prices are important references for all trade between buyers and sellers of fish landings. The crew on board the industrial trawlers and long-liners that land their catches directly to the processing plants are being paid by their share of the landing value calculated to 76 % of the current auction price. This arrangement is part of an agreement with the fisher's union, whose members get their salary as a share of the landed value (Trondsen and Arnarson 2009). The average prices for cod in Iceland in 2000–2008 were in average 2.16 € /kg in the auction market and 1.63 € /kg paid to the industrial landing by trawlers and long-liners in the processing market (Trondsen and Arnarson 2009). The independent fishing vessels selling through auction are thus receiving higher average prices than the average Norwegian cod prices paid to the fishers, whereas the fishers on-board the trawlers and long-liners mainly owned by the processing industry are getting lower prices, which the fishers have accepted because of a higher and more stable catch volume per fisher (Trondsen and Arnarson 2009). The fish channelled through the fish auction value chain are mainly directed to the highest valued fresh fish markets, whereas the industrial fleet delivers catch to the processing value chain. The Icelandic processing industry has, by this system, raw material available at significantly lower cost and higher processing margin compared with their Norwegian competitors.

Fish auctions are also present in the Norwegian cod fisheries, but mainly for fish frozen onboard trawlers. The coastal fleet sells all landings directly to the processing plants protected by the Norwegian Raw Fish Act through the fishers' sales organization minimum price regime. The processing plants in Norway can also purchase raw materials that match their specialized production lines from the frozen fish auctions. However, the frozen fish must be thawed before further processing, which causes water and weight losses, a disadvantage for all except the clipfish industry whose task is to dry water out from the fish (see later discussion). The Norwegian frozen fillets processors therefore have fewer options available than their Icelandic counterparts to cooperate in raw material exchange with their competitors. This limits their ability to cost optimize their production lines to take advantage of the premium prices for fresh raw materials. The Norwegian value chain, lacking the fresh fish auction mechanism, may be characterized as more production-oriented, where standardized products such as H&G fresh and frozen, clipfish, wet salted and stockfish are exported mainly for further processing, sorting and packaging closer to final consumer markets. The key strategic conventions focus on low cost flexible processing of large landed fish volumes compared with the Icelandic system, which allows market specialization through a value chain linked directly to the flexible fresh auction and another linked to the vertical integrated firms.

The main difference in the value chain structures between Norway and Iceland as illustrated in Fig. 9.4, comes from the differences in the two countries' legal regulation and governance conventions covering the first-hand sales of fish. The Norwegian Raw Fish Act gives fisher-controlled organizations authorized by the Ministry of Fisheries monopolies of the first hand sale of all fish landings. At present, four such fisher sales organizations are in practice. Each of these has implemented different sales methods.

Norges Sildesalgslag (for herring fish) and Møre og Romsdal Fiskesalgslag (for cod-fish) in Southern Norway auction whole boatloads of fresh fish. Norges Råfisklag in Northern Norway, the biggest for cod-fish, auctions only frozen fish and sells fresh fish through a regulated minimum price system.

The Norwegian Raw Fish Act requires that all processors that want to purchase fish directly from vessels also have to purchase the entire landed boat load, not only the specific type of fish they want, as they may in fish auctions. In geographically limited areas with few buyers, the consequence has been that the vessels seldom obtain a price higher than the minimum prices set by the fishers' sales organization. Freezing on board the vessel and later storage of the catches in freezing "hotels" for later auction sales opened a new value chain path for the vessel owners to expose all qualities of caught fish to more specialized markets. The introduction of such a frozen auction mechanism has increased the average landing prices above the minimum prices of the fish landed frozen compared with the fresh fish landings. In the period 1997–2003, the price differences between cod sold in Norwegian frozen cod auctions and direct fresh fish landings to processors were between 15 % and 35 % on an annual basis (Helstad et al. 2005). This auction market for frozen fish combined with lower operating costs by extending the trawlers' trip length have became a strong incentive for freezing the catches on-board the vessels. The main buyers of the frozen fish have been the clipfish industry located in Northwest Norway, far away from the fishing grounds in the Barents Sea outside Northern Norway. The clipfish industry took advantage of purchasing frozen H&G fish directly from the fishing companies and bypassing their traditional suppliers of wet salted fish located in Northern Norway close to the fishing grounds. It was a win–win situation for the freezing trawlers and the clipfish industry; cutting off the intermediate wet salted processors from the value chain increased the fish value for the vessels and lowered the clipfish processors' purchase costs compared with what they had to pay the wet salt fish processors. Therefore, these changes strengthened the position of the clipfish value chain in the raw fish market.

The fishers' sales organization, Norges Råfisklag in Northern Norway, decides minimum prices and, unlike the fish auctions in Iceland, does not differentiate prices according to fish qualities, which means that the fishers are paid the same minimum price for higher and lower quality products as long as the quality is acceptable for human consumption. The production conventions for salt fish downstream the value chain have also adapted to the supply profile upstream characterized by low quality differentiation, favouring export of lower quality less sorted wet salted fish and clipfish products. Loss of water through the raw fish freezing process was also an advantage to the clipfish industry where the purpose is to dry out water from the raw material in contrast to the frozen fillet industry that wants to minimize the loss of water in the processing line.

The mixed salt fish products from the primary processors have to be sorted and differentiated further downstream in the value chain to satisfy the demand in specialized end market segments in the consumption regions. These sorting, cutting and packaging operations add value attractive for the Spanish importers of Norwegian unsorted fish. In the case of Iceland, there is a trend to add more of this value to the fish in Iceland before exported. This has become possible by changes in marketing and production management conventions. Information about new demand trends and need opportunities are picked up in market networks and turned into higher value added products and services. The fish auctions that set prices daily for each quality of products maintain the competitive pressure on value adding sorting and processing.

9.2.5 Conventions and Performance

The actual performance figures show that Norway and Iceland are following two different business conventions in the cod export value chain. The export convention applied in the Icelandic value chain is more market-oriented than the more production-oriented Norwegian convention. Different regulation conventions facilitate the differences in industrial orientation and performance that are forming the paths for business transaction in the value chain. Nevertheless, the analysis does not show that the average export value per kilo of the total exported quantity of cod is significantly different; on average, it is only 2 % in Iceland's favour. The presented data cannot prove whether this extra export value is related to differences in the recovery rate from live fish to product weight, or to the higher cost of production that may follow more value adding processing. However if the export values of total landed quantity are divided, we have identified that Iceland has obtained 10 % higher average cod values than Norway in the 2000s compared to 2 % in the 1990s, which indicates Iceland's increasing market oriented value adding position. Two uncertainties have to be taken into account: the conversion factors applied and the amount of imported Russian raw materials in Norway that encounter higher import duties in the EU, which subsequently put downward pressure on the export value. Table 9.3 shows that the calculated export in live weight from Norway has been 10 % higher and Iceland up to 20 % higher than the recorded landings. In addition, about 10–12 % of Norwegian cod landings are consumed domestically in Norway. These figures indicate landing of unregistered catches or that the Norwegian official conversion factors between product weight and live weight are not mirroring the reality. The increase in the Icelandic exported product weight relative to landings of 20 % in the 2000s may indicate the use of additives that increase the water content relative to the official conversion factors in the exported fish.

The imported Russian fish, which account for 27–30 % of the exported cod, may also have reduced the average Norwegian export value because a significant part goes into Norwegian transit and is exported round-frozen unprocessed. The higher EU import duties on Russian fish may also represent a lower net value of exported products of Russian origin compared with the Norwegian and Icelandic fish entering the EU market duty free.

However, even if the higher Icelandic export values represent extra processing costs, it is a contribution to the nation's Net Domestic Product. Nevertheless, the difference is too small to be a significant factor determining which of the two business conventions delivers the highest added value from landings of the total available cod quotas.

The value that can be added is, according to Barney (1996), dependent on the supplied customer value and product rareness, the possibilities for product imitation and the control over a unique supply chain organization (VRIO). If the Norwegian export loses in the competition with Iceland regarding the supplied VRI (customer value, rareness, imitational threat) values, the figures indicate that its high market value relies on control over a unique supply chain of limited and demanded codfish resources. This resource supply control has a strong price commanding power, a resource rent, which partly compensates for the lack of market power generated from a strong and distinctive market orientation and customer satisfaction, where Iceland has the advantage. In fact, Helstad et al. (2005) found that monthly cod prices from 1997 to 2007 at fish auctions in Norway, Iceland and Scotland were significantly co-integrated. This means that, when prices in the Icelandic auction increase, the European importers will look for and buy marginally cheaper alternatives in Norway, which causes a levelling out of significant price differences between identical products. The process may also work the other way around, where Norway's controlling resource power influences the general price level, but buyers who are willing to pay extra for more market prepared products and services direct their demand to Iceland. Menezes et al. (2002) found that the Norwegian ex-vessel prices was the driver of both the Portuguese retail import prices of wet salted cod products from Norway during 1988–1999. Xie and Myrland (2010) also found that the light salted frozen fillets in the Spanish market in the 2000s were a strong substitute for Norwegian salted whole cod. They also found strong substitution effects between the export prices of salted whole cod from Norway, Iceland and the Faroe Islands.

The figures presented show that Iceland commands higher prices in many market segments compared with Norway by supplying differentiated and customer adapted products and services. An important driving force is the Icelandic marketing conventions, which emphasize product sorting and differentiation through fish auctions where all product qualities fetch a specific daily price (Trondsen and Young 2006). Nevertheless, even if high prices are taken out of the big fish sizes in some markets, the big landing volumes of smaller fish still have to go into the processing of lower valued products, which pulls down the average export values. Codfish markets prefer thick fish pieces, made from big spawning cod, harvested mainly in the winter season by the coastal fleet using nets, long-lines and jigging gear. The smaller fish mainly taken by trawlers in the Barents Sea, have a lower size selection in their nets, fishing beyond six nautical miles offshore. An earlier study of the first-hand prices in Norway and Iceland in 1999 showed average prices of big cod (> 3.8 kg) were 65 % higher than the prices of small cod (< 1.5 kg) in fresh fish auctions in northwest Norway, located in the main processing area for clipfish. The similar price premium for big fish in Northern Norway auctions were 54 % and in the Icelandic auctions 47 %. The same study also showed that the landed share of the big cod was very similar in the Icelandic auction (37 %) and in the Northern Norway auction (39 %) (Trondsen et al. 2003). The allocation of fishing rights and quotas between different fleets, seasons and fishing grounds is thus a very important factor influencing fish size and through this the average export value of cod.

In supply chains where many fish sizes, qualities and fish species are flowing, the industry is not free to focus only on the highest paying market segments that require big fish; it also has to have a market strategy for taking care of all landed fish qualities. Icelandic exports of light salted small cod fillets processed in the traditional filleting freezing plants supplied by their own trawlers are a good example of a successful introduction into the Spanish market. However, the market expansion of this low priced cod product pushed the average Icelandic export value per kilo in Spain down to the same level as the Norwegian export value even if the Icelandic exports have been able to take out premium prices for the thick salted cod products in other market segments.

The main structural difference between the two countries' cod value chains is that a larger share of the value added in Norway by minimum price regulations is channelled into the industrial fishers' and vessel owners' pockets compared with Iceland where a larger share is channelled into the processing and exporting industry (Fig. 9.4). The Norwegian industrial fishers are protected by the same minimum prices as the coastal fishermen, even when those industrial vessels are owned by the same company that owns the processing plant. This means that the processing and export industries in Norway work on lower margins compared with Iceland with favourable lower raw material prices regulated to 76 % of the fish auction prices. The independent coastal fishing vessels selling their catches in the Iceland fish auction are obtaining higher prices for fresh quality fish according to consumer market preferences compared with similar vessels in Norway and to Icelandic industrial trawlers. The coastal vessels in Norway are, however, receiving lower prices than the Norwegian industrial vessels (big trawlers and long-liners) because of the common minimum price system and the frozen fish auction opportunity given to the bigger vessels with freezing capabilities (Trondsen et al. 2003). Such value allocation may be a critical constraint on the processing industry's capability to invest in market-oriented value adding activities.

The Icelandic fish processing industry has a relative stronger international presence, which includes daughter offices in the main market areas and more people employed in fish processing compared with Norway. Financial resources are important to enable investment in market-oriented production and product development. Lower Norwegian processing margins may therefore inhibit investment in market-oriented product development and marketing and favour cost-oriented production conventions. On the other hand, Norway has strengthened the generic marketing of Norwegian seafood through Seafood Norway, financed by an export levy. A lack of market-oriented product development conventions in Norway is to some degree compensated by investment in generic marketing.

The Norwegian clipfish industry is a very cost-efficient processing industry supported by cost-oriented industrial conventions and generic marketing. This industry was built up under export and sales monopoly laws after World War II (Hallenstvedt 1982). The laws that were abolished in the 1980s gave a few companies the exclusive right to sell salted cod from the primary processor to the exporter and to actually export the salted cod. This gave the selected firms, mainly located in Northwestern

Norway, far away from the fishing grounds, a protected value chain position from which to build customer loyalty to the main Portuguese and Brazilian markets without disturbing domestic competition. This previously legally protected value chain position was further strengthened after the frozen fish auctions improved the raw material supply independent of location, which has made it difficult for newcomers to enter this value chain after the legal restrictions were abolished. Why did Iceland not enter the clipfish value chain segments? An explanation given by Icelandic industry actors was that the Norwegians were too strong in this chain. When the clipfish chain was closed to them, Iceland chose to expand in other value chains such as light salted codfish for the Spanish market (Trondsen and Arnarson 2009).

Another constraint on market-oriented conventions is the relative stability of raw fish flow through the supply chain. Harvesting and landing of wild fish strongly influenced by biological fluctuations both seasonal and annual, enforce variation in supply pressure throughout the supply chain that the chain managers have to deal with. Both Norway and Iceland are blessed with large migrations of spawning and feeding cod that are present very close to the coast for part of the year. These fish migrations lay the natural basis for seasonal cost-effective fisheries. In the cod-fisheries, the best harvesting season is the winter when the Arctic cod migrate to the fishing grounds for spawning in Lofoten and Western Finnmark. Later, in spring, the cod migrate after capelin to feed in the coastal areas of Eastern Finnmark. This seasonality of fish landings limits the production and value chain choices available to processors and marketers. Salting and drying fish from the winter fishing season for further processing later has traditionally been a production and storage buffer to match seasonal supply 1 year to the main consuming season for salted and dried fish (bacalao) in the next year's Lent season in southern Europe and Brazil, regions dominated by the Catholic religion.

The main reason for the build-up of an industrial distant-cod trawler fleet after World War II was to even out landings for the growing frozen fillet industry independent of the local fishing seasons. The consequences have been more landings of smaller fish and more costly harvesting when fish are spreading out in the feeding areas. Recently, this industry has met strong global competitive pressure from other small frozen fillets made from white fish species including Alaskan pollock from the North Pacific and pangasius from Vietnam, which have caused significant downsizing of the Norwegian frozen fillet industry. The basic natural variation in seasonal landings is still a major force driving supply, which has reduced the scope of production and marketing choices upstream in the cod value chain. Year to year variation in supply is also important. These variations did fit very well into the flexible capacity of the salt fish and clipfish industries, which were able to handle these large variations in landed fish.

Icelandic cod supplies, without any contribution from Russian landings, have been much more stable over the years compared with Norway. A constant supply creates more stability for the long-term development of market-oriented conventions in the industry to increase the value added from the limited landed quantity. The Norwegian processors made their money by taking care of variations in the huge landing quantities. Variation in landings was, at the same time, a disincentive to increasing value added because it reduced the certainty needed for long-term investment in market-oriented product development.

When the landings in Norway were falling after 1998, both the Russian and Norwegian vessels started to freeze fish on-board, which provided more freedom for their sales process to add value by sorting and differentiating many fish qualities in their catches according to specialized customer wants.

9.3 Conclusion and Implications

This paper has shown that market opportunities for increasing added value per kilo of fish can lead to the development of more market-oriented conventions in the value chain. However, it has also shown that there are strong structural forces maintaining path dependencies that strengthen the traditional production conventions constraining market orientation development. In Norway, legal regulation of first hand sales of fish protects the traditional production conventions favouring the traditional industrial fishing sector at the expense of a market-oriented processing sector. The minimum price system has given Norway's fishing fleet higher prices compared with the similar Icelandic fleet, but the Icelandic coastal vessels have access to fresh fish auctions, which give them higher prices than the average Norwegian coastal vessels. The minimum price system also gives the Norwegian fishing plants higher raw material costs compared with their Icelandic competitors. The higher processing and export margin in Iceland has improved the internal value chain resources available for further development of market-oriented conventions. Iceland's success formula appears to be a combination of fresh fish auction markets favouring the highest priced fresh fish export that trades off short-term uncertainty for transaction cost advantage and the vertically integrated hierarchical firms processing value adding products that enjoy the transaction cost advantages of long-term strategies. This formula has allowed Iceland to take advantage of demand in many differentiated market channels.

These structural differences between the Norwegian and the Icelandic supply chain conventions have developed over a long period and are protected by regulation conventions that limit the room for manoeuvre between the strategic harvesting and processing groups.

Our study also shows that on average there were no significant differences in the average total export value per kilo exported or per kilo landings of cod between Norway and Iceland in the 1990s when the effect of Russian landings and re-exports from Norway are taken into account. However, our calculations indicate 10 % higher Icelandic export value per kg landings compared with Norway in the 2000s. The Norwegian market position reflects its control over a scarce, in-demand natural resource creating a resource rent, and probably the generic marketing effort of the Norwegian Seafood Export Council. However, the study shows value adding opportunities at the product market level. Turning the market opportunities into value adding management conventions is constrained by the industry's management environment embedded in regulatory conventions.

The main elements in the Norwegian locking-in mechanism are the minimum price conventions, managed by Norges Råfisklag in Northern Norway where the biggest cod fisheries are located, which do not differentiate prices according to actual market prices for different product qualities to the same degree as Iceland does. In this region there is no fresh fish auction system serving the coastal vessels, only one for frozen fish serving industrial vessels with on-board freezing. In the Northwest Norway, where the clipfish processing industry dominates, first hand sales are organized through a fresh and frozen fish auction. These first hand sales supply structures give price incentives to frozen fish landings in north and north western Norway, which is the opposite to Iceland where all raw cod-fish are landed fresh.

Our study is consistent with conventional neoclassical economic theory that markets open to all interested entrepreneurs and organized like the fresh fish auction markets in Iceland and the frozen fish market in Norway are a very efficient allocation mechanism. This mechanism is both motivation for and realization of value adding both directly and indirectly by imposing competitive pressure to transform scarce heterogeneous resources in supply regions into homogeneous products matching market wishes in consumption regions. These market structures are limited by the organization and flexibility of the value chain. This might be described as a train travelling through a tunnel. Each business manager sees only the customers in the carriage ahead, which represents the outlet for the production, the supplier carriage behind, which represents the inputs of raw materials and the walls, roof and rails in the tunnel which represent the structures created by nature and regulations restricting the market area in which trading takes place freely. When consumer and value chain market preferences and demand are changing, market-oriented entrepreneurs can only respond by adapting products and services if they are able to link together a chain of carriages on rails all the way back to the catching sector to satisfy and take advantage of the market changes before competitors do so. Coordination of each value adding carriage activity in the entire value chain between catch and consumer requires business conventions regulating accepted practice among participants in the chain. When such traditional government-protected conventions emerge in one period and are confronted by new market conditions demanding new marketing conventions, the traditional conventions become constraints on entrepreneurial creation of new value chains to adapt to new market trends. In most industries, competitive pressure from emerging new entrepreneurs forge new value chains from alternative sources in globalized markets. This forces the traditional value chains in the market arena to change or die. In our case, of harvesting resource rents from a limited natural resource, this is not necessarily the case.

This study illustrates that harvesting and processing of a scarce fish resource can be very competitive even when relying on traditional cost-oriented production conventions and partly ignoring the changes taking place in the consumer market while focusing the business conventions on tapping the rent from the resource. It is an open question whether entrepreneurs in such an industry are given the opportunity to link together new trains of value chain carriages required to bring the demanded products to markets. Two factors are important. (1) Is the supply chain open to entrepreneurial expansion from both incumbents and new entrants? (2) Is there interest in the industrial environment in entering the industry? In fisheries, the fish quota allocation system represents a very high entry barrier for new entrepreneurs, which favours the incumbents while the first hand sales regulation system favours the industrial catching sector in competition to add value from the fish landings. The second question must be judged given the alternatives facing the entrepreneurs in an industry or country. In Norway, with its very low unemployment rate, high labour costs and a high resource rent in simply harvesting the resource, investors will not necessarily look for opportunities in a labour intensive industry such as fish processing. Therefore, it is more likely that entrepreneurs in other countries with a high unemployment rate, such as Spain, will discover and exploit market opportunities for value adding based on raw materials processed by cost-oriented Norwegian raw material suppliers.

References

Barney, JA (1996) Gaining and Sustaining Competitive Advantage. Addison-Wesley, New York

Bendiksen BI (2004) Driftsundersøkelsen i fiskeindustrien – Oppsummering av inntjening og lønnsomhet i 2003. Report 14/2004. Fiskeriforskning AS, Tromsø, Norway

Bendiksen BI (2010) Driftsundersøkelsen i fiskeindustrien. Oppsummering av lønnsomheten i norsk fiskeindustri i 2008. Report 11/2010. Nofima AS, Tromsø, Norway

Dicken P (1998) Global shift: transforming the world economy. Paul Chapman, London

Hallenstvedt A (1982) Med lov og organisasjon: organisering av interesser og markeder i norsk fiskerinæring. Universitetsforlaget, Oslo

Helstad K, Vassdal T, Trondsen T (2005) Price links between auction and direct sales of fresh and frozen fish in North Norway (1997–2003). Mar Res Econ 20:305–322

Lindkvist KB, Gallart-Jornet L, Stabell MC (2008) The restructuring of the Spanish salted fish market. Canadian Geographer 52(1):105–120

Menezes R, Dias JF, Guia F, Filipe JC, Guerreiro V (2002). European markets integration: the case of the Norway–Portugal cod value chain. Conference paper, XIV European Association of Fishery Economists, University of Algarve, Faroe, Portugal, 25–27 March

Porter ME (1990) The competitive advantage of Nations. Macmillan, New York

Salais R, Storper M (1992) The four worlds of contemporary industry. Camb J Econ 16:169–193

Statistics Iceland. (2010) http://www.statice.is/Statistics/Fisheries-and-agriculture/Financial-Accounts. Accessed 15 Dec 2010

Trondsen T (1995) Product salt cod, Proceedings of the 7th Biennial Conference of the International Institute of Fisheries and Trade, Taipei, Taiwan, Republic of China, July 18–21, 1994, vol. 3, pp. 97–118

Trondsen T, Johnston RS (1998) Market orientation and raw material control. J Market Focused Management 3:193–210

Trondsen T, Young JA (2006) The role of fish auctions in value adding in fish marketing chains. In: Ashe F (ed) Primary industries facing global markets. The supply chain for Norwegian food. Universitetsforlaget, Oslo, pp 393–427

Trondsen T, Arnarson I (2009). Salted fish processing in Iceland. A field survey report. Working paper, Norges Fiskerihøgskole, University of Tromsø

Trondsen T, Helstad K, Young JA (2003) Market-oriented regional fisheries management—an analysis of four fish regions in the North Atlantic. Ocean Coast Manag 46:917–941

Wathne KH, Heide JB (2004) Relationship governance in supply chain network. J Mark 68:73–89

Xie J, Myrland Ø (2010) Modeling market structure of the Spanish salted fish market. Scand J Food Econ 7:119–127

Chapter 10
Challenges of the Norwegian Salted Fish Industry in the Spanish Market

Jinghua Xie and Øystein Myrland

Abstract The Spanish market for salted fish products has been restructured over the last decade. A significant pattern is a steady increase in the consumption of frozen light salted fillets at the expense of traditional products. The three Nordic countries, (i.e., Norway, Iceland and the Faroe Islands) account for around 90 % of the Spanish market share. Among them, the share of Iceland is expanding, due to the substantial growth of light salted fillet export. The share of Faroe Islands slightly increases due to its success in clipfish fillet exporting in consumer packages. However, the Norwegian export suffers a major loss, which is mainly explained by its conventional commitment of export of wet salted cod. Overall, the results indicate a significant challenge for the Norwegian salt fish industry to maintain its position in the Spanish market. A market-oriented strategy of differentiating and developing new products is suggested to improve the Norwegian industry's performance in Spain.

Keywords Salted fish trade · Market structure · Norwegian market problems · Econometric modeling · Frozen light salted fillets · Spain

10.1 Introduction

The product group, salted fish, in the Spanish market consists of three main traditional products and one relatively new product called frozen light salted fillets. The traditional ones are whole wet salted cod, salted cod fillets and clipfish. Wet salted cod is a product where the gutted fresh fish is split and salted. Salted fillets are wet salted cod fillets, which are boneless and without a belly. Clipfish is a wet salted cod dried to around 70–75 % of the wet salted weight. Light salted fillets are the frozen fillets either soaked or injected with 2 % salt, which tastes similar to the desalted traditional salted fillets[1].

[1] For a more detailed description of the products, see Lindkvist and Sanchez (2008).

J. Xie (✉) · Ø. Myrland
School of Business and Economics, University of Tromso, Tromso, Norway
e-mail: Xie.jinghua@uit.no

K. B. Lindkvist, T. Trondsen (eds.), *Nordic-Iberian Cod Value Chains*,
MARE Publication Series 8, DOI 10.1007/978-3-319-16405-2_10

In Chap. 3, Richter-Hanssen analyzed the historical background of the Norwegian-Spanish salted fish trade and generalized the following points. The history of this trade dates back to 1665, and clipfish was the only product in the early period. The Spanish market had never been a large market for Norwegian salted fish until 1970, when Spain's own production tremendously decreased. In recent decades, the share of salted whole fish and fillets has been increasing, while clipfish has been decreasing to a marginal percent. The Spanish consumers' complaints about Norwegian salted fish originated back to the 1790s. Richter-Hanssen also concluded "Norwegians, however, had been on the defensive and proud of an old, entrenched position."

The focus of our analysis in this chapter is the modern situation of Norwegian salted fish in Spain. The chapter includes four sections. Section 10.1 briefly discusses the data we have employed. Section 10.2 presents the general situation of Spanish imports of salted fish and problems for the Norwegian salted fish industry. Section 10.3 discusses the estimated results of a more rigorous demand model, which confirm the market situation depicted by the descriptive data in Sect. 10.2. Finally, Section 10.4 suggests a market-oriented strategy for the Norwegian salted fish industry to improve its performance in Spain.

10.2 Data

The data sources for our analysis are monthly trade data of Eurostat and national statistics of Norway, Iceland, and the Faroe Islands, all provided by the Norwegian Seafood Council (NSC).

The original data were specified according to product description and the associated HS number for each product. We aggregated the data according to our needs. Specifically, to keep the data consistent with that in the other chapters of the book, in Sect. 10.2, the quantity of salted fish products is measured by product weight, otherwise noted. However, in Sect. 10.3, to make the econometric modeling manageable, the aggregation of different product forms becomes necessary. It demands that the quantity be measured in REW (Round-Fish-Equivalent-Weight). The quantity on the REW basis for Norwegian export is available from the original data set given by the NSC. Those for Iceland and the Faroe Islands are converted from product weights according to the corresponding conversion factors given by the Norwegian Ministry of Fisheries. The values are measured in euros at the FOB level. Unit prices were computed by dividing the value by quantity.

We found that Eurostat is not consistent with the national data for the specific commodity, (i.e., for the commodity with the same HS number, the Norwegian export to Spain is not consistent with the Spanish import from Norway). The reason for this occurrence, as explained by the Norwegian Seafood Council, might be that Eurostat registered the commodity according to the first landing countries in EU countries, while the Norwegian data are registered according to who is the buyer in the contract. It is probably the same for the Icelandic and Faroese data. This means the EU data on Spanish imports from EU countries might also include some commodities that are merely passing through the country for transportation reasons. Therefore, we use Eurostat for the data of total Spanish imports and use the data of these three nations as their exports to Spain, respectively.

The ambiguity of Eurostat and exporting countries' data is a problem for researchers making an empirical analysis. Fortunately, for our analysis, the problem is not serious since the three Nordic countries make up 90 % of the total Spanish imports and we use the relatively aggregated data.

Another problem with the data is that thus far there is no registered record of light salted cod fillets either in Eurostat of Spanish imports, or in the Icelandic export data until the year 2008; whereas, since the mid-1990s, the brine and injecting salt solutions have changed the content of frozen fillets. Although no one wants to talk about it, everyone knows about it and the market seems to be happy with the development (Lindkvist et al. 2008). This claim is strongly confirmed by the Icelandic data in 2008, when for the first time, Iceland had light salted fillets in its statistics. According to the Icelandic statistics of 2008, the exports of frozen fillets to Spain were 1113 t, and the exports of new recorded frozen light salted fillets were 5790 t, for a total of 6905 t. The relative shares of frozen fillets and frozen light salted fillets indicate that at least 80 % of the frozen fillets from Iceland before 2008 were actually light salted. Therefore, we use the export data of frozen fillets from Iceland as a proxy for Spanish imports of frozen light salted fillets.

Although the real data of the light salted fillets is available in 2008, however to keep the data consistent, we added the frozen fillets and frozen light salted fillets for the year 2008. There might be some light salted fillets exported by the Faroe Islands too. While, according to an interview with some importers published by Lindkvist et al. (2008), the majority of them are from Iceland. In addition, there are still no available data for the Faroe Islands to indicate how many frozen light salted fillets are exported to Spain. Thus, when we mention the Spanish imports of light salted cod hereafter, we actually use the data of the Icelandic exports of frozen fillets based on the above-discussed reasons.

The total Spanish imports from other countries (e.g., China, Canada, and Russia) comprise about 10 %. To give the addressed problem manageable dimensions, the study therefore focuses on salted fish from the three Nordic countries.

10.3 Descriptive Comparative Analysis of the Spanish Salted Fish Market and the Problem for the Norwegian Industry

10.3.1 Spanish Market of Salted Fish

The total Spanish imports of salted fish grew rapidly between the years 1991 and 2007 (Fig. 10.1). The import volume (in product weight)[2] doubled from 25,000–50,800 t. Among those, the traditional wet salted whole fish and fillets grew by 25 % and clipfish by 283 %. The main contribution was from frozen light salted fillets, which

[2] Hereafter in product weight, otherwise noted.

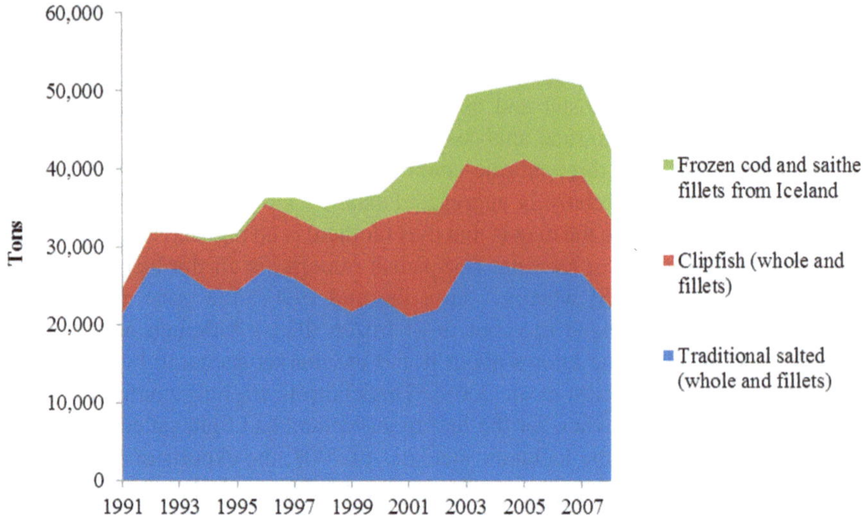

Fig. 10.1 Spanish imports of salted fish. (Source: NSC (2009))

grew from 47 to 11,532 t in this period. As a result, the market share of salted whole fish and fillets dropped from 86 to 52 %, clipfish grew from 13 to 25 %, and frozen light salted fillets grew from almost 0–23 %. This pattern held in the year 2008, although the imported amounts of all the products declined due to the world economic crisis.

For traditional salted whole fish and fillets, the export amounts of Norway, Iceland, and the Faroe Islands account for about 90 % of Spanish imports (Fig. 10.2). The main difference exists for clipfish. The three Nordic countries only account for 15–20 % of the Spanish imports, and the remainder is shared between four EU countries, namely Denmark, the Netherlands, Portugal, and France. Whether this is because of the different statistic criteria between Eurostat and the national statistics of the exporting countries or because a large amount of clipfish is processed in these EU countries or both is not clear. However, this may indicate that, compared with traditional salted whole fish and fillets, more clipfish is processed in the EU countries and re-exported to Spain.

Spanish imports of salted fish include cod, saithe, haddock, ling, blue ling, and others. Cod has the dominate share of more than 90 %; some other white fish species, however, seem to be taking up more shares and keeping the Spanish market growing (Fig. 10.3). Eurostat aggregates all the other species together. To see what the main species are that are beginning to appear in the Spanish salted fish market, we have graphed the Nordic exports of different species to Spain for the period 1991–2008 in Fig. 10.4. It shows that saithe is the main contributor: the export of saithe increased from 707 to 2303 t in the period. Ling makes a minor contribution. The exports increased from 785 to 1516 t in the period. The exports of all the other species are still comparatively tiny. The saithe exports are mainly as a product of frozen light

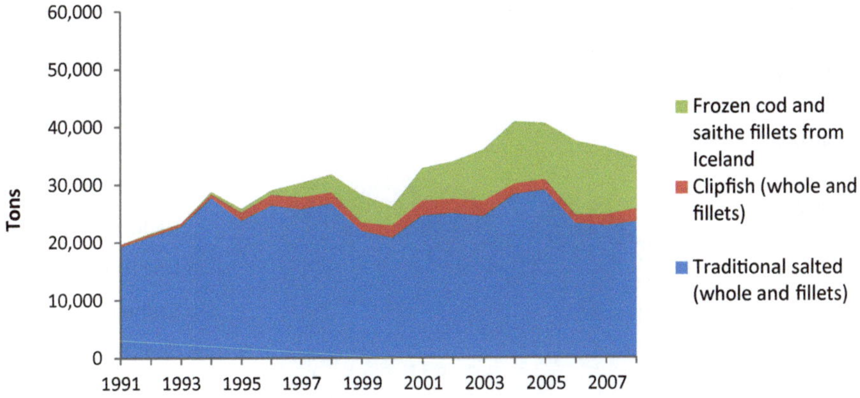

Fig. 10.2 Nordic exports of salted fish to Spain. (Source: NSC (2009))

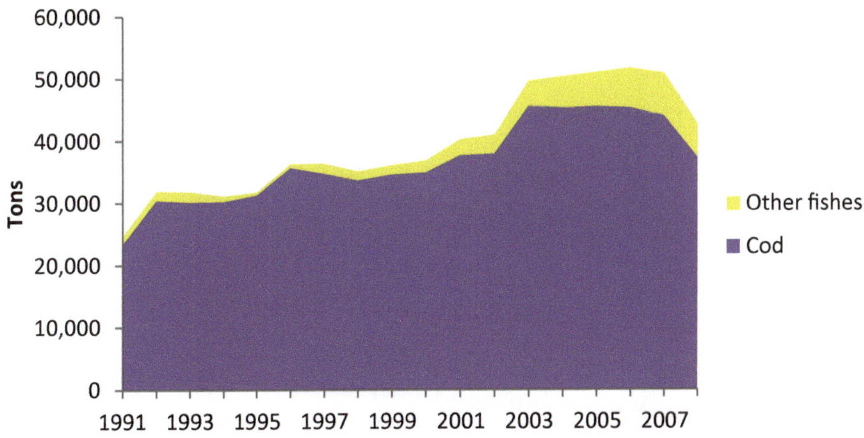

Fig. 10.3 Spanish imports of salted fish. (Source: NSC (2009))

fillets from Iceland to Spain. The ling exports are in the form of traditional salted whole fish and fillets.

Figure 10.5 shows that the growth of the Spanish market generally is not equally shared by Norway, Iceland, and the Faroe Islands. In the early time of the market (1991), Iceland had the dominant share of 65 % in the salted cod market, followed by the Faroe Islands with 21 %, and Norway with 14 %. After that, the Norwegian industry seemed to try to expand the Spanish market. In 1998, the total Norwegian exports of salted cod reached 11,500 t, accounting for 45 % of the Spanish market. After that, the exports stagnated at the amount of 5000–7000 t with a market share around 15–25 %. On the other hand, Icelandic exports steadily increased from 11,600 to 18,700 t between 1991 and 2007. The share of Iceland in recent years was around 60 %. The exports of the Faroe Islands were rather stable, 4000–8000 t between the years, with a market share from 15 to 30 %.

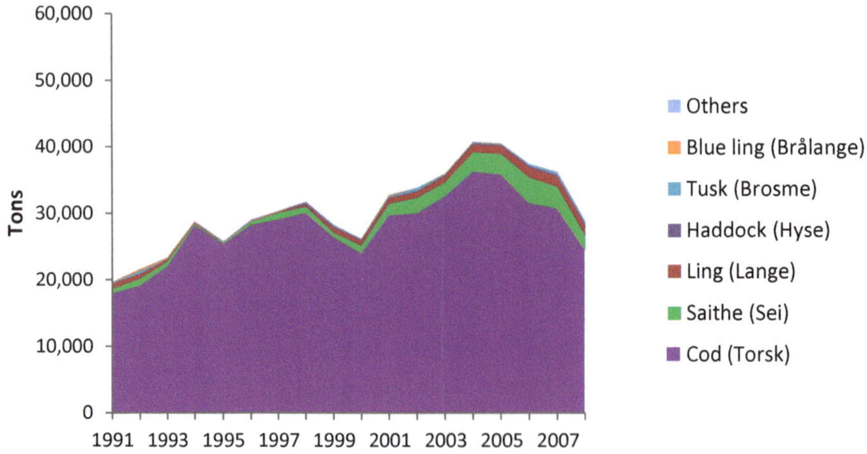

Fig. 10.4 Nordic exports of salted fish to Spain. (Source: NSC (2009))

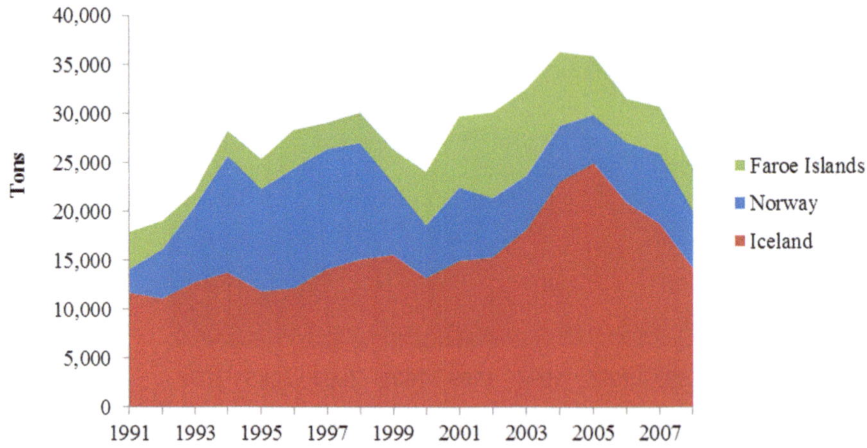

Fig. 10.5 Spanish imports of salted cod from Nordic countries. (Source: NSC (2009))

10.3.2 Problems of the Norwegian Salted Fish Industry in the Spanish Market

Exports of salted fish are generally decided by both the import demand for the product and the supply of raw fish. According to Trondsen (1994), all North Atlantic cod fisheries in the EU countries, Norway, Iceland, the Faroe Islands, Greenland and Canada have been through a process where the cod quotas have been reduced year by year. Thus far only regulations of the cod stock in Norway have been successful. The Norwegian cod stock has recovered, and the quota and catch again increased from 1991. Therefore, the reason for the loss of the Norwegian market share to Icelandic salted cod in Spain seems primarily to be a direct result of poorer marketing performance of the Norwegian salted fish industry.

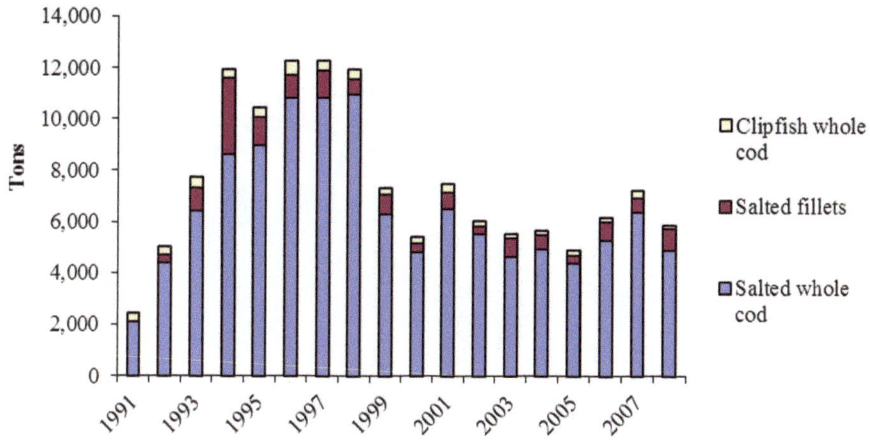

Fig. 10.6 Norwegian exports of salted cod to Spain. (Source: NSC (2009))

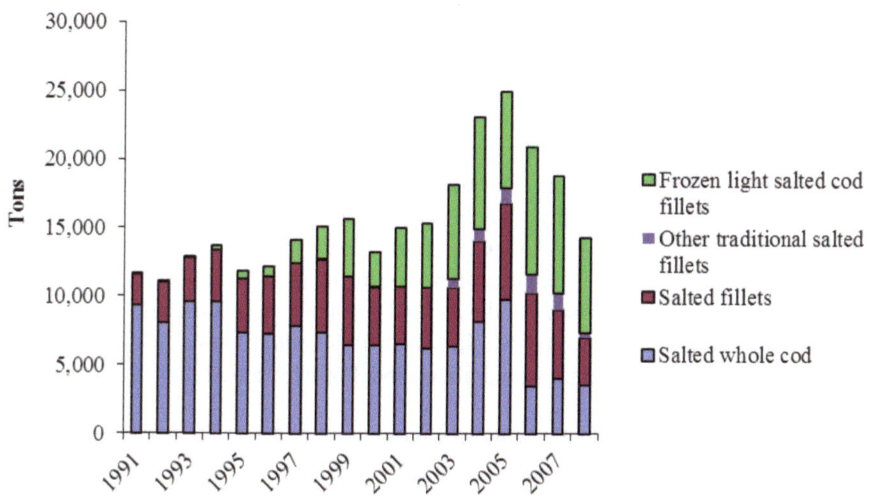

Fig. 10.7 Icelandic exports of salted cod to Spain. (Source: NSC (2009))

According to standard marketing and strategy literature, value growth of commodity products is related to industries' ability to segment buyer preferences, differentiate, and focus their product marketing strategies towards the most attractive buyers (Porter 1980). Figure 10.6 shows that Norwegian exports to Spain stick to salted whole fish, which dominated 85% of its total exports throughout the years from 1991 until recent years. Opposite to Norway, Iceland has many more product differences. From 1991 to 2007, it cut the exports of salted whole fish from 9300 to 4000 t, slightly increased the exports of fillets from 2300 to 5000 t, and greatly expanded the exports of frozen light salted fillets from 47 to 8600 t (Fig. 10.7). In 2007, it had a market distribution of 25 % for salted whole fish, 24 % for salted fillets and 46 % for

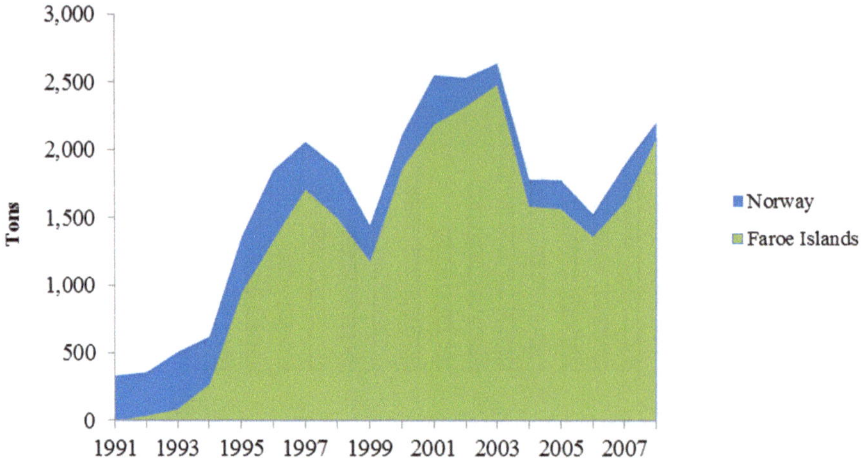

Fig. 10.8 Spanish imports of clipfish from Nordic countries. (Source: NSC (2009))

frozen salted fillets. In addition, it also tried to export salted cod eggs and tongues: 6 % in 2007.

If we compare how many product items specified as salted cod are exported to Spain in the national statistics of Norway, Iceland, and the Faroe Islands, there are 3 for Norway, 15 for Iceland, and 4 for the Faroe Islands. Table 10.1[3] presents that the 3 items from Norway are salted whole fish, salted fillets, and clipfish. No product forms or packages have been changed since the year 1990, which is the earliest year in our data set. The reality is that it has never been changed since 1980 when the large scale salted fish export started. For the Iceland product group, it includes 14 different products, 5 of which are salted whole fish or light salted whole fish in consumer packages, 2 of which are light salted fillets, and 3 of which are other salted cod such as tongues, belly, and eggs.

What we can conclude from these figures is that the Icelandic industry tried different product forms and packages to differentiate their products and to satisfy consumers in different segments. Although the Faroe Islands have only four items, the majority of their clipfish fillets are in consumer packaging. This more convenient packaging for consumers might be one of the reasons that the Faroe Islands has dominated the market share of clipfish in Spain (Fig. 10.8).

As discussed above, Spanish imports of traditional wet salted whole fish and fillets dropped from 86 to 52 % in the last 16 years, and the growth of the Spanish market was mainly a result of import of frozen light salted fillets. Since the market for traditional products is significantly shrinking and the Norwegian industry still keeps producing and exporting traditional products, it is no surprise that Norway is losing market share.

[3] According to the product HS numbers listed in the national statistics of these countries.

Table 10.1 Products exported by Norway and Iceland to Spain

Norway		Iceland	
Wet salted whole cod			
1990–2009.03	03056200 Cod, salted	*1990–2006*	03056209 Cod, other salted
		2007	03056299 Cod, other salted
			03056294 Cod, flattened, consumer packaging, salted
		2008–2009.03	03056299 Cod, other salted
			03056219 Cod, in consumer packaging, other salted
			03056212 Cod, in consumer packaging, light salted parts with skin and bone
			03056211 Cod, in consumer packaging, light salted parts without skin and bone
			03056214 Cod, in consumer packaging, other salted parts with skin and bone
			03056213 Cod, in consumer packaging, other salted parts without skin and bone
			03056294 Cod, flattened, in consumer packaging, salted
Salted cod fillets			
1990–2009.03	03053004 Cod, salted fillet	*1990–2007*	03053011 Cod, salted fillet
		2008–2009.03	03053032 Cod, salted fillet, otherwise
Clipfish			
1990–2008	03055107 Cod, clipfish	*2008–2009.03*	03053022 Cod, light salted frozen fillet, otherwise
2009.01–2009.03	03055108 Cod, Atlantic clipfish		03053012 Cod, light salted frozen fillet in consumer packaging
	03055109 Cod, Greenland/pacific clipfish		

The problem is even worse for the Norwegian industry if we look at the prices of different products from different sources (Fig. 10.9). To make prices comparable, the prices of the products in Fig. 10.9 are calculated in round fish weight. It shows that prices of the three Norwegian products are the second cheapest only next to the Icelandic light salted fillet. Norwegian salted whole cod, the dominant product in Norwegian export, is much cheaper than its main competitors, salted whole cod from Iceland and the Faroe Islands.

When the traditional salted fish market reaches maturity, products from different suppliers substitute each other, and prices become the only competitive tools. Businesses have, according to Porter (1980), three strategic options: cost cutting,

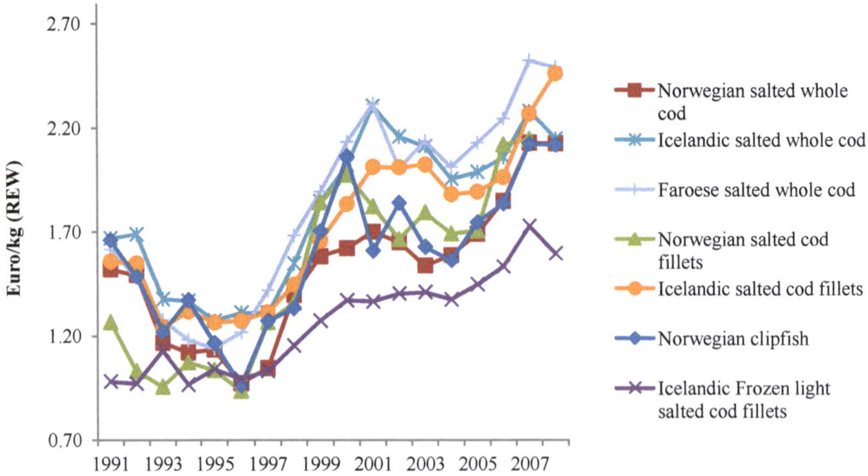

Fig. 10.9 Prices of Nordic exports of salted fish to Spain. (Source: NSC (2009))

differentiation, and focus, or a combination of these strategies (Trondsen 1994). However, when the Norwegian products are much lower than that of competitors and production of salted fish largely depends on raw fish, it is questionable how much room there exists for the Norwegian salted fish industry to lower its prices.

To summarize, the new growing exports of frozen light salted fillets from Iceland are a strong substitute for traditional salted products, while Norway is dominant in export of traditional wet salted whole cod. In the relatively even shared wet salted whole cod market, products from these three Nordic countries are expected to strongly compete against each other. The price of Norwegian salted whole cod is much cheaper than that of its competitors. The strategy of lowering price to expand market share is not feasible in the long run. We therefore conclude that the Norwegian industry is facing a great challenge in the Spanish market. In the next section of econometric modeling, the estimated results of econometric modeling confirm these market characters.

10.4 Econometric Modeling of the Spanish Salted Fish Market

Xie and Myrland (2010) estimated the demand for salted fish commodities in Spain using the Almost Ideal Demand System (AIDS) and specified it as:

$$w_{i,t} = \varphi_i + \sum_{k=2}^{4} \phi_{ik} D_{k,t} + \varsigma_i T_t + t_i T_t^{\,2} + \sum_{j=1}^{6} \sigma_{ij} \ln p_{j,t} + \theta_i \ln(X_t/P_t) + \lambda_i W_{i,t-1} + u_{i,t},$$

$$i = 1, \ldots, N; \quad t = 1, \ldots, T.$$

where $N = 6$ is the number of salted fish products in the system, and i indexes the equation for each product. $w_i = p_i q_i / \sum_{j-1}^{6} p_j q_j$ is the budget share of the ith good, and $\ln P = \sum_i w_i \ln p_i$ is the Stone price index. $X = \sum_{j-1}^{6} p_j q_j$ is the total Spanish import value of salted fish products in the system. D_k is the seasonal dummies and T is the trend.

The estimated results suggest that a structural change may be at work in the Spanish salted fish market. This claim is based on estimated parameters of the trend variable T presented in Table 10.2. The positive sign of the trend parameter of a product indicated that keeping prices and total expenditure on salted products constant, consumers will buy more of this product, which is explained by the change of consumers' preference. Therefore, the positive signs of Icelandic light salted frozen fillet, clipfish and Faroe Island salted whole equations in Table 10.2 suggest that consumer preferences for these products are strengthening. Paralleling, negative signs of Norwegian salted whole cod and Icelandic salted whole cod indicate consumers' preference for these two products are weakening. This result therefore confirms the finding in Sect. 10.3, that the Spanish market is restructured by a significant increase in consumption of frozen light salted fillets at the expense of traditional products. The estimated parameter of T^2 suggests the speed of the trend. For example, the positive sign of T^2 in the Norwegian equation means that the consumer's preference for buying less Norwegian salmon increases.

The estimated Hicksian price elasticities in Table 10.3 give us an insight into the relative strength of substitution relationships. The positive and large magnitude of e_{13}, e_{21} and e_{32}, indicate that salted whole fish from different sources are competing strongly against each other[4]. This result is expected since the Spanish market for salted whole cod is a traditional saturated market, relatively on average shared by Norway, Iceland, and the Faroe Islands.

The cross price elasticity of Norwegian salted whole fish with respect to frozen light salted fillets is 1.09. This means that a 1 % decline of light salted fillet prices will drag down the Norwegian salted whole fish demand by 1.09 %. Icelandic light salted fillets are thus a strong substitute for Norwegian salted whole fish.

10.5 Suggestions for a Market Strategy for the Norwegian Salted Fish Industry in Spain

The problems for Norwegian salted fish in Spain are basically concluded from the comparative analysis of the Spanish market in Sect. 10.3 and the stricter econometric analysis of the Spanish market structure in Sect. 10.4. Norway is losing competence in both quality and quantity in the Spanish traditional market and failing to adapt production to Spains growing demand for new products and new packages. In agreement with Lindkvist (2009), the overall suggestion for the Norwegian salted fish industry to improve the situation is to shift from a production-oriented strategy to

[4] For a detailed discussion, see Xie and Myrland (2010).

Table 10.2 Estimates results of the trend variables (sample 94.01–09.02). (Source: Edited based on table IV in Xie and Myrland (2010))

Independent variables	Norway Salted whole fish	Iceland Salted whole fish	Faroe Islands Salted whole fish	Nordic Salted fillets	Nordic Clipfish	Icelandic Light salted fillets
T	−0.003	−0.001	0.002	−0.001	0.001	0.002
	(−5.12)*	(−2.31)*	(4.09)*	(−1.36)	(3.21)*	(5.45)*
T^2	0.00001	0.00000	−0.00001	0.000003	0.00000	−0.000002
	(3.19)*	(−0.002)	(−3.21)*	(1.49)	(−2.83)*	(−0.90)

Numbers in parentheses are asymptotic t-ratios, *, ** indicate significance at the 5 and 10 % levels, respectively

Table 10.3 Estimated Hicksian price elasticities (long run). (Source: Edited based on table VII in Xie and Myrland (2010))

Quantity demanded from	ei1	ei2	ei3	ei4	ei5	ei6
Norwegian salted whole cod	− 1.13	− 0.63	0.75	0.27	− 0.35	1.09
	(− 3.27)*	(− 1.45)	(2.77)*	(0.62)	(− 1.24)	(2.71)*
Icelandic salted whole cod	0.56	− 0.50	0.02	0.59	0.07	− 0.75
	(2.21)*	(− 1.56)	(0.08)	(1.84)**	(0.35)	(− 2.53)*
Faroe Island salted whole cod	− 0.24	1.49	− 1.00	0.69	0.52	− 1.46
	(− 0.58)	(2.85)*	(− 3.05)*	(1.30)	(1.52)	(− 3.01)*
Salted cod fillet	0.27	0.15	0.01	− 0.97	0.16	0.37
	(1.37)	(0.61)	(0.05)	(− 3.84)*	(0.98)	(1.63)**
Clipfish	− 0.20	0.04	0.26	1.18	− 0.64	− 0.65
	(− 0.62)	(0.10)	(1.03)	(2.88)*	(− 1.70)**	(− 1.72)**
Frozen light salted cod fillet	0.51	0.13	− 0.44	− 1.10	0.06	0.84
	(1.45)	(0.29)	(− 1.59)	(− 2.46)*	(0.15)	(2.05)*

Numbers in parentheses are asymptotic t-ratios, *, ** indicate significance at the 5 and 10 % levels, respectively a market-oriented strategy. From the market point of view, the suggestion can be further specified as follows.

10.5.1 Improve the Quality of Fish

Figure 10.9 in Sect. 10.3 indicates that, for the same product categories, the prices of Norwegian products are uniformly much cheaper than those from Iceland and the Faroe Islands. In addition, low prices are associated with small market shares. This confirms the findings given by Lindkvist (2009) from a panel study between 1998 and 2006 that Spanish importers consider salted fish from Iceland and the Faroe Islands to be of better quality than that of Norwegian products. It is also consistent with the historic findings of Richter-Hanssen in Chap. 3: The Norwegian salted fish industry has never succeeded in improving quality according to Spanish complaints in the long term. With a growth of income, people do not eat food simply to fill their stomachs. They are willing to pay for quality. As pointed out by Lindkvist (2009), single actors meet barriers against deliveries of the same products as their competitors in the Norwegian control system regulating the production chain. It is time for government, organizations and industries to work together to solve the problem, given that the whole industry is facing the same problem.

10.5.2 Increase the Exports of Salted White Fish Other than Cod

Figure 10.10 and also Fig. 10.4 in Sect. 10.3 show that, over the last decade, Spain has expanded imports of other salted white fish, mainly saithe and ling. Its market share grew from 6 to 14 % between 1991 and 2009. Although the expansion speed is not that fast, the pattern of growth is always present without any exception between the years. Figure 10.10 suggests that the Norwegian industry started quite early in exploring this market. However, the market was taken up by Iceland and the Faroe Island in recent years. Again, the price of Norwegian fish is much lower than that of its competitors.

10.5.3 Enhance the Product Variety

It is not unrealistic to assume that consumer demand for food is actually the demand for the attributes of food, which include nutrients, taste, smell, appearance, safety, culture etc. (Lancaster 1966). The choice of attributes is constrained by prices of products, disposal income of the consumer, and consumer preferences. Social factors such as size of the family, level of education, age of the head-of-household and so on are important driving factors (Cheng and Capps 1988; Asche 1996; Kinnucan et al. 1993; Cortez and Senauer 1996; Myrland et al. 2000; Olsen 2003; Altintzoglou 2012). Therefore, the demand for food products must be various and vibrant. The point for producers is to provide various products to meet the different demands from each subgroup.

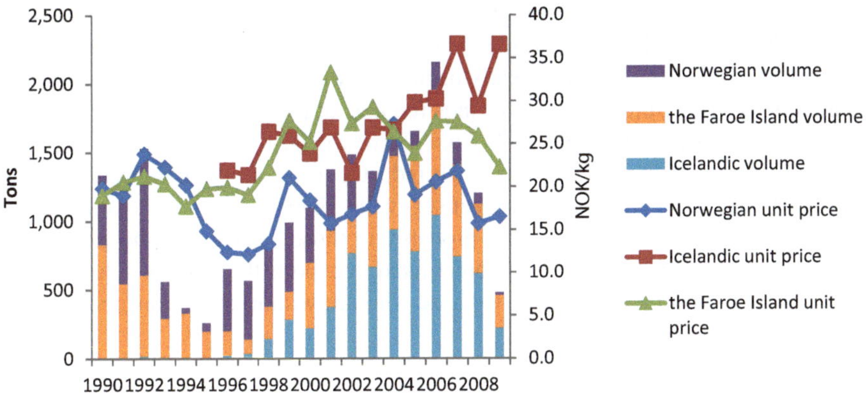

Fig. 10.10 Nordic exports of other salted fish (not cod) to Spain. (Source: NSC (2009))

Norway has been providing the exact same products of salted whole cod and salted cod fillets to Spain. The only change is that, starting in 2009, it separated clipfish according to the sources of catches, (i.e. Atlantic and Greenland clipfish, respectively). In contrast, Iceland provided seven different salted whole cod, three different salted cod fillets, and two different frozen salted light salted fillets starting in 2008. The Faroe Island provided the clipfish in consumer packages. Their variety happened both in the product itself and the packaging of products. Market success depends on many conditions, but the basic principle is that producers should adapt their production to market needs. When the market's needs become more sophisticated, it is necessary for producers to come up with a greater variety of products.

10.5.4 Segment the Market According to the Need of Each Customer Subgroup

As mentioned, consumers' demands for the attributes of products are constrained by economic factors such as prices and income and also by social factors such as size of the family, education level, and so on. Therefore, the demand is a function of prices, income, and social factors. We can find a common characteristic of a subgroup that prefers to buy a specific product. For example, busy young couples might prefer to buy frozen light salted fillets with consumer packaging since it is easier to cook, while retired people might prefer to buy high quality traditional salted cod since they have time to cook, and it also provides a good memory of the past (Geeroms et al. 2008). The importance is to segment the market and provide the products according to the needs of specific sub-markets.

As pointed out by Lindkvist (2009), market conventions are not homogenous. Although new light salted products have encroached on the Spanish market, traditional salted fish still accounted for about 50 % of the market share. A good example is the Faroe Islands' positioning of selling salted whole cod to the customers who prefer the traditional product with good quality. Although the total Spanish demand for salted whole fish has been greatly decreasing over the last decade, the export of salted whole fish from the Faroe Islands remains at the same level and is the most expensive among all the salted cods from Norway (Fig. 10.9).

The success of the Icelandic salt fish industry in Spain is not entirely but very primarily due to its attention to market segments. Figure 10.9 shows that it exports the cheap frozen light salted fillets probably to those consumers who prefer fast, tasty, and cheap salted fish. On the other hand, it exports the salted whole fish with and without consumer packaging to those who are willing to pay for good quality conventional salted fish.

10.6 Conclusion

The salted fish products in Spain consist of three main traditional products including whole wet salted cod, slated cod fillets and clipfish, and one relatively new product called frozen light salted fillets. The total Spanish imports of salted fish grew rapidly between the years 1991 and 2007. This is mainly attributed to a huge expansion of frozen light salted fillets. Both the data description and the econometric results indicate that the Norwegian salted fish industry is losing competence to the Icelandic industry in the Spanish salted fish market. The main problem of the Norwegian industry is failing to adapt to Spain's growing demand for new products and new packages. Similar to Lindkvist (2009), our overall suggestion for the Norwegian salted fish industry is to shift from a production-oriented strategy to a market-oriented strategy. Specifically, they are: to improve the quality of fish, to increase the exports of salted white fish other than cod, to enhance the product variety and to segment the market according to the needs of customer subgroups.

References

Altintzoglou T, Sveinsdottir K, Einarsdottir G, Schelvis R, Luten JB (2012) Evaluating of seafood product concepts by young adults and families with young children from Denmark, Norway and Iceland. J Aquat Food Prod Technol 21:418–432

Asche F (1996) A system approach to the demand for salmon in the European Union. Appl Econ 28:97–101

Cheng H, Capps O Jr (1988) Demand analysis of fresh and frozen finfish and shellfish in the United States. Am J Agric Econ 70(3):533–542

Cortez R, Senauer B (1996) Taste changes in the demand for food by demographic groups in the United States: a nonparametric empirical analysis. Am J Agric Econ 78:280–289

Geeroms N, Verbeke W, Kenhove PV (2008) Consumers' health-related motive orientations and ready meal consumption behavior. Appetite 51:704–712

Kinnucan HW, Nelsen RG, Hiariey J (1993) U.S. preferences for fish and seafood: an evoked set analysis. Marine Res Econ 8:273–291

Lancaster KL (1966) A new approach to consumer theory. J Polit Econ 74(2):132–157

Lindkvist KB (2009) Innovations and market response in the Norwegian salted fish industry. Project report of the Norwegian-Spanish salted fish project. University of Bergen

Lindkvist KB, Sanchez JL (2008) Conventions and innovation: a comparison of two localized natural resource-based industries. Reg Stud 42(3):343–354

Lindkvist KB, Lorena GJ, Stabell MC (2008) The restructuring of the Spanish salted fish market. Can Geogr 52:105–120

Myrland O, Trondsen T, Johnston RS, Lund E (2000) Determinants of seafood consumption in Norway: lifestyle, revealed preferences, and barriers to consumption. Food Qual Prefer 11:169–188

Norwegian Seafood Council (NSC) (2009) Personal contact with Kristin Lien, market analyst at Norwegian Seafood Council. http://en.seafood.no/. Accessed Mar 2009

Olsen SO (2003) Understanding the relationship between age and seafood consumption: the mediating role of attitude, health involvement and convenience. Food Qual Prefer 14:199–209

Porter ME (1980) Competitive strategy: techniques for analyzing industries and competitors. The Free Press, New York

Trondsen T (1994) Product marketing strategies and performance in the European market for salted ground fish. In: Proceeding in International Institute of Fisheries Economics and Trade (IIFET), pp. 134–142

Xie J, Myrland O (2010) Modeling market structure of the Spanish salted fish market. Food Econ-Acta Agric Scand Sect C 7:119–127

Chapter 11
Market Power of the Icelandic Salted Fish Industry in Spanish Markets

Jinghua Xie

Abstract Iceland and Norway are two main salted fish exporters to Spain. An evident result is that the Icelandic industry benefits from a growing Spanish salted fish market at the cost of the Norwegian industry. The market-oriented innovation of Icelandic industry mainly explains its success against production-oriented Norwegian industry. I argue in this chapter that Icelandic innovation is co-developing with the market power of the Icelandic industry, which makes the Norwegian problem even worse. The chapter tests the existence of Icelandic market power in Spain by using descriptive data analysis and econometric modeling. The sources of Icelandic market power were stated to be friendly Icelandic institutional environments, the industry's market-oriented innovations, and the control power of the Icelandic salted fish industry in the vertical integrated salted fish value chain.

Keywords Salted fish · Econometric results of market power · Market power and innovation · Spain · Iceland · Norway

11.1 Introduction

Norway and Iceland are the world's two most important Atlantic Cod suppliers. Around 40 % of the cod catches in these two countries are processed to salted cod and exported mainly to the European Union (EU). Due to both culture and religion, Spain has a proud tradition of salted cod consumption. It is the second largest salted fish importer in the EU, next only to Portugal.

Both Norway and Iceland have a long history of salted cod trade with Spain. The history of Norwegian and Spanish salted fish trade goes back to 1665, while Iceland was dominant in the Spanish market in 1952 (Richter-Hanssen, Chap. 3). In recent decades, the Spanish market has undertaken a fundamental restructuring (Lindkvist et al. 2008). The total Spanish imports of salted cod doubled from 25,000 to 50,800 t between 1991 and 2007. Significant changes happened in both the product

J. Xie (✉)
School of Business and Economics, University of Tromso, Tromso, Norway
e-mail: xie.jinghua@uit.no

K. B. Lindkvist, T. Trondsen (eds.), *Nordic-Iberian Cod Value Chains*,
MARE Publication Series 8, DOI 10.1007/978-3-319-16405-2_11

structure and quality image of the fish. Spanish consumers began to buy more frozen light-salted cod, newly introduced by the Icelandic industry, and the convenient ready-to-cook products in consumer packages (Xie and Myrland 2010). Spanish consumers also displayed a growing demand for white, thick, and juicy fish rather than thinner, dry, yellowish fish (Lindkvist et al. 2008).

In the face of new challenges from the Spanish market, the Norwegian salted fish industry seems to have lost its market to the Icelandic industry. Xie and Myrland, Trondsen, Lindkvist and Sanchez-Hernandez have discussed the problems of the Norwegian salted fish industry in Spain in several chapters in this book. Xie and Myrland find that there is a strong substitute of light salted cod exported by Icelandic industry to the traditional Norwegian salted whole cod. Trondsen (Chaps. 9 and 13), Lindkvist (Chap. 7), and Sanchez-Hernandez (Chap. 4) argue that the legal regulative, the goals of the innovation system and the other structural characteristics make the Norwegian salted fish industry sea-based conventions, or, in other words, production-oriented conventions. They suggest that the mismatch between producers' conventions and consumers' demands is the core reason preventing the Norwegian industry from taking benefit from a significant increase of the Spanish salted fish demand. Although the other chapters in this book have conducted analyses of the Spanish salted fish market, no one has estimated and quantified the competition between the two biggest suppliers, Iceland and Norway, directly.

The focus of this chapter is the market power of the Icelandic salted fish industry in competition with the Norwegian industry in Spain. I argue that the problems of the Norwegian salted fish industry discussed by Trondsen, Lindkvist and Sanchez-Hernandez partly provide an opportunity for the Icelandic industry to develop its market power in the Spanish market. The Icelandic market power, on the other hand, worsens the problem of the Norwegian industry.

The chapter is organized as follows: It starts with a short section on methodology. Next, some signals of the Icelandic market power are discussed, and an econometric modeling is used to test the Icelandic market power. Sources of the Icelandic market power and co-development of innovation and market power are then discussed. The chapter ends with a brief conclusion.

11.2 Methodology

The data used in this chapter came from several sources. The monthly data of quantity and value of salted cod exported from Norway and Iceland between 2000.01 and 2009.02 were provided by the Norwegian Seafood Council (NSC). The data were updated to December 2009. The updated data were from Statistics Norway and Statistics Iceland. The product weight in the data was converted to REW (Round-Fish-Equivalent-Weight). The conversion factors were given by the Norwegian Ministry of Fisheries. The unit prices were computed by dividing the value by quantity on REW. The imputed prices were converted to the importing countries' currency (euro) using the exchange rate data obtained from a website developed by Professor Werner Antweiler (Antweiler 2010). The landing price, the wage for fish processing and the interest rate data of Norway and Iceland were from the state statistics of each country. The real income and the harmonized index of consumer prices (HICP) data of Spain were obtained from Eurostat website (Eurostat 2010). The original wage data of Norway and Iceland and the income data of Spain are quarterly. They are converted to monthly data using the cubic spline method.

Icelandic market power in Spain is first measured by the Icelandic market share and the price difference between the Icelandic and Norwegian product. A more rigorous econometric measurement is then conducted by estimating residue demand elasticity (RDE) models of the Icelandic and Norwegian salted fish exports. Goldberg and Knetter (1999) measured the intensity of competition that a group of exporters face in a particular export market by the residual demand elasticity. The RDE model was applied widely in studies of international agricultural trade (Carter et al. 1999; Poosiripinyo and Reed 2005; Song 2006).

11.3 Signals of the Icelandic Market Power in Spain

Market power is conventionally measured by market share and the Lerner index, defined as $L = (P - MC)/P$ where P is output price and MC is the marginal cost of output (Lerner 1934). In a competitive market, a firm's $MC = P$ and the firm has no extra profit. When a firm has market power, it then has the ability to mark up its price above its marginal cost. The definition of Lerner's index suggests that when $L = 0$ there is no market power; the bigger number of L indicates the higher market power of a firm. The Lerner index can therefore be used to measure the market power of a firm.

Goldberg and Knetter (1999) questioned measuring market power by market share. They explained that a firm with a dominant market share may still be constrained in its market power if the demand for the product has substantial substitutes or if the supply of competing firms is elastic. And likewise, a firm with a small market share may be able to have its market power if it plays in a market niche of highly differentiated products. To compute the Lerner index, as defined above, we need the marginal cost of output, which is extremely difficult to measure. Until now, there has been no solid literature to address the problem, which makes the use of the Lerner index limited.

In this study, the Icelandic market power in Spain is both measured by the Icelandic market share and a proxy of the Lerner index. Although the measurement of market share has its drawbacks as discussed by Goldberg and Knetter (1999), market size is anyhow a good indicator of market power. In most cases, it is a big firm or industry that has the ability to control the market, not a small one. Moreover, because of the simplicity, market share is widely used by the governments of many countries in the world to infer market power of a firm in antitrust cases. For example, to limit industry concentration, Norwegian law stipulates that no single firm can hold more than 25 % of the salmon production licenses.

To get the marginal cost of the Icelandic salted fish is near impossible. However, a comparison of the Icelandic and Norwegian salted fish prices can be a good proxy of the Icelandic Lerner index. Norway and Iceland both are wealthy northern European countries. They have almost the same cod resources and other production costs. Moreover, landing price of raw fish is 7–11 % higher in Norway as compared to Iceland (Trondsen, Chap. 9). Thus, if the Icelandic salted fish price is higher than the Norwegian price, it must be higher than its own marginal cost ($L > 0$), and the Icelandic industry has market power in the Spanish market.

Table 11.1 presents the export of both Norway and Iceland to the Spanish market in 1994 and 2008. Norway had dominant export of traditional wet salted whole cod both in 1994 and 2008. However, the export volume of the traditional wet salted cod decreased from 8656 to 4886 t between 1994 and 2008. At the same time, its export of traditional salted cod fillet and clipfish decreased from 2934 to 882 t and from 349 to 119 t, respectively. There was no export of frozen light salted cod fillets from Norway in either period. As a result, the Norwegian market share in Spain decreased from 42 to 24 % between 1994 and 2008. On the other hand, Icelandic exports were more evenly distributed in the different product categories. The traditional wet salted cod exports decreased from 9618 to 3539 t. Traditional salted fillets export slightly decreased from 3727 to 3422 t. The frozen light salted cod fillets grew tremendously from 0 to 6905 t and account for around 50 % of the total Icelandic exports to Spain. As a result, the total Icelandic exports grew from 49 to 58 % between 1994 and 2008. It is evident that the Spanish market was relatively evenly shared by Norwegian and Icelandic industry in 1994, while it was more dominated by Iceland in 2008.

Table 11.1 shows the same result given by Trondsen (Chap. 13) that the Icelandic industry has not entered the clipfish market and left it to the Norwegian industry. Similarly, the Norwegians do not enter the frozen light salted cod market due to the Norwegian legal barrier of using phosphate in salted fish production. Clipfish is a wet salted cod dried to around 70–75 % of the wet salted weight. Light salted fillets are frozen fillets either soaked or injected with 2 % salt. Both clipfish and light salted cod fillets are quite different from wet salted cod. To make the prices of the products comparable, we aggregated the wet salted whole cod and fillets based on REW (Round-Fish-Equivalent-Weight).

The prices used in Fig. 11.1 are FOB prices in euro. It depicts that the Icelandic prices of the traditional wet salted cod were uniformly higher than the Norwegian between 2001 and 2009. As discussed earlier, the export prices of Norwegian cod should be, under any circumstances, higher than the marginal cost of Icelandic salted cod. Icelandic prices in Fig. 11.1 is averagely 20 % higher than the Norwegian, which means the sign of the Lerner Index is positive ($L > 0$) and the Icelandic industry has market power in the Spanish market.

Table 11.1 Norwegian and Icelandic export of salted cod to Spain. (Data source: NSC 2009)

Exporters	Norway	Iceland	Norway	Iceland
	1994 volume (1000 kg REW)		*2008 volume (1000 kg REW)*	
Salted whole cod	8656	9618	4886	3539
Salted cod fillets	2934	3727	882	3422
Clipfish	349	–	119	–
Frozen light salted cod fillets	–	–	–	6905
All	11,939	13,345	5887	13,866
	1994 market share		*2008 market share*	
Salted whole cod	0.42	0.47	0.46	0.33
Salted cod fillets	0.44	0.56	0.20	0.80
Clipfish	1.00	0.00	1.00	0.00
Frozen light salted cod fillets	0.00	1.00	0.00	1.00
All	0.42	0.49	0.24	0.58

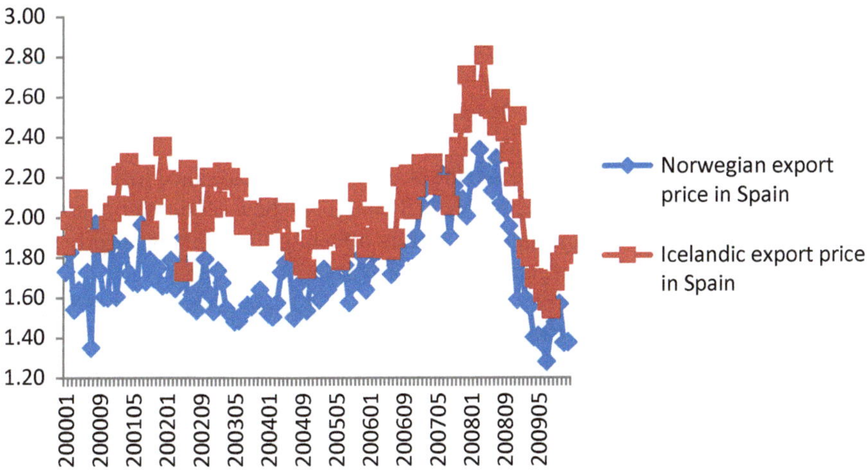

Fig. 11.1 Export prices of wet salted cod from Norway and Iceland to Spain. (Data sources: NSC (2009); Statistics Norway 2010; Statistics Iceland 2010 and Statistics Faroe Islands 2010)

11.4 Econometric Results of Market Power of the Icelandic Salted Fish Industry in Spain

Both the Icelandic market size and proxy of the Lerner Index above indicate the market power of the Icelandic salted fish industry in Spain. To get statistical evidence, we estimated the following residue demand model based on the theoretical framework developed by Goldberg and Knetter (1999). Table 11.1 evidently indicates market power in Iceland's trade of light salted frozen cod fillets and in Norway's clipfish trade. Due to no supply of these two products from each of the competitor, the two countries have market powers in each niche market, which emerge from product differentiation. The competition of the two countries in the traditional markets of wet salted cod (whole fish and fillets) is estimated in the following model.

The empirical model estimated for the Icelandic residual demand in Spain:

$$\begin{aligned}\ln EXP_{I,t} =& \alpha_0 + \alpha_1 \ln EXQ_{I,t} + \alpha_2 \ln LP_{N,t} + \alpha_3 \ln E_{N,t} + \alpha_4 \ln INT_{N,t} + \alpha_5 \ln W_{N,t} \\ & + \alpha_6 \ln LP_{F,t} + \alpha_7 \ln E_{F,t} + \alpha_8 \ln W_{F,t} + \alpha_9 \ln I_{S,t} + \alpha_{10} \ln HICP_{S,t} \\ & + \alpha_{11} D_t + \lambda_I \ln EXP_{I,t-1} + \sum_{m=2}^{12} \theta_m Month_{m,t}\end{aligned}$$

where EXP_I and EXQ_I are the export price and quantity of Iceland to Spain, respectively. *LP, E, INT* and *W* are the landing price, exchange rate, interest rate, and wage, respectively.

The suffix *N* and *F* index Norway and the Faroe Islands respectively. I_s is the disposable consumer income of Spain. *HICPs* is the harmonized index of prices in Spain. A dummy variable *D* is included to explain the collapse of salted cod prices after the Economic Crisis started from mid-2008. The lag dependent variable $EXP_{I,t-1}$ is incorporated, assuming it takes longer than 1 month for an exporter to adjust fully to changes in prices. $Month_m$ is included to capture the seasonal effect on the demand for salted cod.

Although both the market share and the prices of Norwegian salted fish indicate no market power of the Norwegian industry in Spain, to see how econometric modeling works and also get an insight into the factors affecting the Norwegian prices, we estimate the following Norwegian equation also as:

$$\begin{aligned}\ln EXP_{N,t} =& \beta_0 + \beta_1 \ln EXQ_{N,t} + \beta_2 \ln LP_{I,t} + \beta_3 \ln E_{I,t} + \beta_4 \ln INT_{I,t} + \beta_5 \ln W_{I,t} \\ & + \beta_6 \ln LP_{F,t} + \beta_7 \ln E_{F,t} + \beta_8 \ln W_{F,t} + \beta_9 \ln I_{S,t} + \beta_{10} \ln HICP_{S,t} \\ & + \beta_{11} D_t + \lambda_N \ln EXP_{N,t-1} + \sum_{M=2}^{12} \vartheta_m Month_{m,t}\end{aligned}$$

Similarly to that in the Icelandic equation, EXP_{N} and EXQ_N are the Norwegian export price and quantity to Spain, respectively. The definitions of the other variables are the same as that in the Icelandic equation.

LP_F, E_F, and W_F are incorporated in both Icelandic and Norwegian equations as the production costs of the other competitor except for Norwegian or Icelandic industry in the Spanish market. No Faroese interest rate variable is included due to unavailable data. I_s and $HICP_s$ are included in both equations to explain the effect of consumer demand on the export prices of salted cod from Norway and Iceland.

Laws of supply and demand suggest that price and quantity correspond simultaneously with the long term market clearing equilibrium. The explanatory variables of quantity *EXQ* are therefore endogenous in both Icelandic and Norwegian equations.

The two-stage least square (2SLS) estimation is thus used to estimate the models. The cost variables of Iceland or Norway including the landing price of the cod, the interest rate, the wage cost, and the exchange rates are used to run *EXQ* in Icelandic or Norwegian equation respectively.

Table 11.2 presents estimated results for the Icelandic and Norwegian residual demand models. Because we are using a double log model, the estimated parameters can be interpreted directly as flexibility[1]. The long-run flexibilities are the short-run flexibilities divided by $(1 - \lambda_{N/I})$. The estimated λ_N is not statistically significant in the Norwegian equation, which means the short-run estimation is equal to the long-run estimation.

Goldberg and Knetter (1999) explain that a significant value of the residual demand flexibility suggests market power[2]. The estimated coefficients of *EXQ* (α_1 and β_1) in our model are statistically significant in the Icelandic equation, but not in the Norwegian equation, providing evidence that the Icelandic salted fish industry has market power in the Spanish market, but the Norwegian industry does not. This result is expected and consistent with what was indicated by the market size and proxy of the Lerner's index analyzed in Sect. 11.3.

The market power analysis suggests that the Spanish salted fish price is largely decided by the Icelandic fish industry. Both the Norwegian and the Faroese industries are price takers, since they are small players in the market. Market power makes it possible for the Icelandic salted fish industry to mark up the price–marginal cost margin ($L > 0$). This also explains why the Icelandic industry can set a higher price than its Norwegian competitor (Fig. 11.1), and enjoy the lion's share of the market at the same time (Table 11.1).

In the Icelandic equation, none of the estimated parameters of either the Norwegian or Faroese cost variables (i.e., raw cod landing price, exchange rate, interest rate, and wage) are significant except for the Norwegian wage cost. On the other hand, five of seven Icelandic and Faroese cost variables in the Norwegian equation are significant. The results again suggest the market power of the Icelandic industry in Spain. The Icelandic prices are largely decided by its own export volume (significant α_1). The supplies of other competitors do not affect its export price significantly. However, Norwegian export prices are not decided by its own quantity (insignificant β_1), but by the Icelandic and Faroese exports. The larger magnitudes of LP_I and LP_F mean that the landing prices of raw cod are of most importance to affect export prices, which is reasonable since costs of raw cod account for around 65 % of the salted fish production. Estimated coefficients of LP_I LP_F are 0.28 and 0.18, respectively, in the Norwegian equation, which means that a 1 % reduction in the Icelandic or Faroese landing price of raw cod will drag down the Norwegian export price to Spain by 0.28 and 0.18 % respectively. The relatively bigger magnitude of the coefficient of LP_I over that of LP_F is explained by market power of the Icelandic industry in Spain.

[1] Flexibility is defined as $f_{ii} = \frac{\partial P_i}{\partial Q_i} \frac{Q_i}{P_i}$ for one's own quantity effect. It is inverse price elasticity, and thus also called elasticity in some literature (e.g., Goldberg and Knetter 1934).

[2] For a detail theoretical discussion, see Goldberg and Knetter (1934).

Table 11.2 Estimated results of residue demand model of Iceland and Norway

Iceland	$\ln EXQ_I$	$\ln LP_N$	$\ln E_N$	$\ln INT_N$	$\ln W_N$	$\ln LP_F$	$\ln E_F$	$\ln W_F$	$\ln I_S$	$\ln HICP_S$	$\ln EXP_{I,t-1}$	D
Short run	− 0.127**	0.02	− 0.25	0.01	− 1.639*	0.10	1.99	− 0.03	− 0.65	3.53*	0.445*	− 0.10**
	(− 1.722)	(0.213)	(− 0.881)	(0.197)	(− 2.249)	(1.202)	(0.319)	(− 0.479)	(− 1.106)	(2.279)	(3.740)	(− 1.825)
Long run	− 0.23**	0.04	− 0.46	0.02	− 2.95*	0.17	3.58	− 0.05	− 1.18	6.36*		− 0.18**
	(− 1.722)	(0.213)	(− 0.881)	(0.197)	(− 2.249)	(1.202)	(0.319)	(− 0.479)	(− 1.106)	(2.279)		(− 1.825)
Norway	$\ln EXQ_N$	$\ln LP_I$	$\ln E_I$	$\ln INT_I$	$\ln W_I$	$\ln LP_F$	$\ln E_F$	$\ln W_F$	$\ln I_S$	$\ln HICP_S$	$\ln EXP_{I,t-1}$	D
	− 0.04	0.28*	0.05	0.07*	− 1.06*	0.18*	1.49	0.06*	0.37	2.22**	0.13	− 0.09
	(− 0.802)	(2.778)	(0.495)	(2.815)	(− 2.155)	(2.635)	(0.218)	(1.192)	(0.534)	(1.730)	(1.155)	(− 1.592)

Numbers in parentheses are asymptotic t-ratios, *, ** indicate significance at the 5 and 10% levels, respectively
R^2 is 0.70 and 0.79 in the Icelandic and Norwegian equation respectively

The estimated coefficients of demand shifter ($HICP_S$) are 3.53 in the short run and 6.36 in the long run for the Icelandic equation, and 2.22 in the Norwegian equation. They are positive and statistically significant. This means that when the prices of other commodities goods increase, the prices of salted cod significantly increase too, and vice versa. This suggests that in the Spanish market, there are strong substitutes for salted cod demand. This result is consistent with the discussion by Trondsen (Chap. 13) and Xie and Myrland (Chap. 10): the cod industry has met hard competitive pressure from other white fish species like saithe, Alaskan pollock, and pangasius from Vietnam.

11.5 Sources of Icelandic Market Power in the Spanish Salted Fish Market

Trondsen (Chap. 9 and 13) has written about the Norwegian individual vessel quota (IVQ) regime that motivates vessels to concentrate their catching in the peak fishing season with a high catch rate. This practice makes it difficult for the Norwegian salted fish processing industry to get consistent quality fish throughout the whole year. However this is possible for the Icelandic industry. Under the Icelandic individual transferable quota (ITQ) system, quotas are owned mainly by vertically integrated companies involving both processing plants and fishing vessels. Furthermore in Norway, since the processors do not control the fish quota, they normally have to make contracts with a number of vessels and agree to buy the entire catch of the vessels contracted, although some of the species landed, and their specifications, may not necessarily fit the ideal product profile of the processors (Trondsen et al. 2003). By contrast, implementation of the fresh fish auction in Iceland ensures the processing industry can select the exact specie, size, and quality of fish that the final markets need. At the same time, the landing price of cod in Norway is protected by a minimum price managed by the Norwegian Fishermen's Organization (*Norges Fiskarlag*), which partly causes the average landing price in Norway to be 7–11 % higher than the corresponding Icelandic price.

Facing the challenge of consumer innovation in demand for convenient and prepared products, industry innovation is evidently necessary. Xie and Myrland (Chap. 10) suggest that Iceland has many more diversified products compared to Norway. Fifteen different products in terms of forms, cuts, sizes, and packages are exported from Iceland to Spain, compared to only 3 from Norway. Increasing employment of Spanish women in a wide variety of jobs, plus the growth of supermarket chains, require sustainable and standard ready-to-go products. The consistent supply of Icelandic salted fish in consumer packages and ready-to-eat light salted fillets perfectly match the innovations of consumer and retail demands in Spain. According to the survey conducted by Lindkvist (Chap. 7), the only dominant convention that all but 2 of the 28 respondents of Norwegian salted fish producers agreed to is the convention of snow-white colour. Espinosa-Segui and Alba (Chap. 5) also mentioned that the thickness convention is recognized in a more or less similar way.

Although the use of phosphates to make salted cod look white, thick, and juicy is considered questionable and is prohibited by EU regulation of food safety, the Icelandic motivation of providing products in line with Spanish consumers' preference is unquestionable.

The Icelandic salted fish value chain is ruled by the Icelandic industry, which coordinates market relationships with fishermen and local Spanish distribution channels. The industry pays the best price for the best quality fish and brings market information downstream to fishing boats. Spain is a complex market with many submarkets based on geographical features (Espinosa-Segui and Alba, Chap. 5). To deal with requests and complaints from retailers and consumers, many of the Icelandic companies have set a network of supporting agencies in Spain. Working as a mediator, the Icelandic industry motivates fishermen to treat the fish well and meet consumers' expectations. Many processing plants in Iceland have their own fishing trawlers and quotas. Such a vertical integration reduces uncertainty of seasonal price and quality volatility and guarantees more homogenous and standardized products. On the other hand, Norwegian relationships with buyers in Spain had been traditionally organized by independent Spanish fish brokers. In the late 1990s, both parties became suspicious of each other. The situation has been gradually improved due to the efforts of both Norwegian industry and the Norwegian Seafood Council (NSC).

Structural changes in the Spanish demand for white, thick cod and convenient products have challenged exporters to adapt their supplies to the new consumer preferences. The Icelandic industry has succeeded over the Norwegian industry in adapting to those changes and accumulated market power in the Spanish market. Based on the above analysis, I conclude that the friendly institutional environments, the industry's market-oriented innovations, and the control power of the value chain are working together to make the Icelandic industry the primary market power in the Spanish salted fish market.

11.6 Co-development of Market Power and Innovation

Innovation, as discussed above, is not the only source, but a key component of what gives Iceland market power in the Spanish salted fish market. Schumpeter (1942), on the other hand, argues that large firms with market power accelerate the rate of innovation. The logic behind this dual approach is that innovation depends on industry being able to appropriate returns from the introduction of new products or services. In a Schumpeterian economy, industry with strong market positions gains more from innovation because market power can stimulate technological progress on the expectation of receiving monopoly rents (Nicholas 2003). The estimated results of market share effect on patent counts (Nicholas 2003) confirm Schumpeter's assumption about the innovation-market power dynamic. However, on the other hand, Schumpeter does not deny the common belief of positive effect of competition on innovation. He explains that market power and competition can have positive effects on innovation concurrently, since industry is always facing the problem of creative destruction.

Larsen et al. (Chap. 2) argues that innovation can be producer pushed and/or consumer pulled. In the case of the Spanish salted fish market, it is important to know that producers not only comply with the needs of consumers, but also try to teach their customers about products. When the producers are powerful and control the value chain, they have more bargaining power to influence consumers' preference. Although the convention of whiteness as a trait of fish quality is pervasive due to the influence of the fresh fish market (Sanchez-Hernandez, Chap. 4), the overwhelming influence of the Icelandic industry might be another reason to establish consumer's image that snow white Icelandic cod is better than the natural golden color of Norwegian cod.

The wine case study in Sanchez-Hernandez (Chap. 4) also suggests that nature, health, countryside, tradition and reputation are fashionable in the Spanish market. Compared to the Icelandic products, which are suspected of using chemical additives such as phosphates, the thin yellowish Norwegian cod is more natural, healthy, and traditional, which should be valued by the Spanish consumers. The survey of Lindkvist (Chap. 7) shows that a majority of consumers will not buy products injected with chemicals, such as phosphates, if they know they have been used. Supposing that Norwegian industry has a strong position in Spain, it is possible for them to reshape the consumers' image of quality fish.

11.7 Conclusions

In this chapter, descriptive analysis of market share, proxy of the Lerner's index and econometric estimation of the Icelandic residue demand model, all suggest the existence of market power owned by the Icelandic industry in the Spanish salted fish market. For Iceland, friendly institutions and relevant government policies have set a favorable climate for Icelandic industry innovation and the possibility of controlling the whole value chain from Icelandic boat to the Spanish table. These factors strengthen each other and work together to assure the Icelandic industry of primary market power in the Spanish salted fish market.

Although the success of the Icelandic industry in the Spanish market is mainly driven by the industry's market-oriented innovation, I argue that market power and innovation is co-developing. Marginal return to innovation is higher when the industry has market power, which is consistent with the result that Icelandic traditional salted whole cod and fillets are 20 % more expensive than the corresponding Norwegian price. Although the perfect match of the Icelandic innovation with Spanish consumers' demand mainly explains the dominant position of Icelandic industry in the growing Spanish market, the Icelandic industry is also using its market power to influence Spanish consumers' preferences as well.

References

Antweiler W (2010) Pacific Exchange Rate Service. The University of British Columbia, Sauder School of Business. http://fx.sauder.ubc.ca/data.html. Accessed 22 July 2010

Carter CA, MacLaren D, Yilmaz A (1999) How competitive is the world wheat market. Working Paper No. 99–002. Department of Agricultural and Resource Economics, University of California Davis

Eurostat (2010) http://epp.eurostat.ec.europa.eu/portal/page/portal/eurostat/home. Accessed 20 July 2010

Goldberg PK, Knetter MM (1999) Measuring the intensity of competition in export markets. J Int Econ 47:27–60

Lerner AP (1934) The concept of monopoly and the measurement of monopoly power. Rev Econ Stud 1(3):157–175

Lindkvist KB, Lorena G-J, Stabell MC (2008) The restructuring of the Spanish salted fish market. Can Geogr 52:105–120

Nicholas T (2003). Why Schumpeter was right: innovation, market power, and creatively destruction in 1920s America. J Econ Hist 63(4):1023–1058

Norwegian Seafood Council (NSC) (2009) Personal contact with Kristin Lien, market analyst at Norwegian Seafood Council. http://en.seafood.no/. Accessed March 2009

Poosiripinyo R, Reed M (2005) Measuring market power in the Japanese chicken meat market. J Int Agric Trade Dev 1:135–148

Schumpeter J (1942) Capitalism socialism and democracy. Harper, New York

Song B (2006) Market power and competitive analysis of China's soybean import market. PhD dissertation. University of Kentucky, USA

Statistics Faroe Island (2010) http://www.hagstova.fo/portal/page/portal/HAGSTOVAN/Statistics_%20Faroe_Islands. Accessed 17 July 2010

Statistics Iceland (2010) http://www.statice.is/. Accessed 15 July 2010

Statistics Norway (2010) http://www.ssb.no/english/. Accessed 15 July 2010

Trondsen T, Helstad K, Young J (2003) Market-oriented regional fisheries management? An analysis of four fish regions in the North Atlantic. Ocean Coast Manage 46(9–10):917–941

Xie J, Myrland Ø (2010) Modeling market structure of the Spanish salted fish market. Food Econ—Acta Agric Scand, Sect C 7:119–127

Chapter 12
Knowledge Conventions and Public Infrastructure

Torbjørn Schei

Abstract This chapter focuses on the contribution to industrial development from educational vocational training (EVT) in a period when the Norwegian fish processing industry has been in decline. Its pool of skilled labor has eroded in particular for the salted fish industry, characterised by a low level of formal competence and no tradition for seeking higher education. This dominating cost-oriented convention in an industry composed of relatively small enterprises has not demanded sufficient specialized, higher educated, people to support market oriented production. Thus an increasing mismatch between knowledge conventions in the fishing and processing industries, and in the more academic oriented development of an education infrastructure, is only now evolving.

Keywords Institutions · Conventions · Subtle alliances · Vocational education · Learning region

12.1 Introduction

Differences in path development of industrial conventions reflect differences in investment and accumulation of specialized knowledge and networks (Barney 1996). The competence and networks needed for a market-oriented production may be different from a cost-efficient production strategy (Porter 1990). This chapter focuses on an important, but much neglected theme in local industrial development, including the salt fish industry, namely the contribution from educational vocational training (EVT). The chapter addresses a possible devastating turn around in EVT in the Norwegian fishing industry, from a leading, vibrant role to one of decline. The chapter inquires into the erosion of a significant external economy of scale—a pool of skilled labor, and explores how a public good and an important external economy is actually constructed and deconstructed. Explanations are sought in learning and knowledge conventions and how these are translated into public governance of an education infrastructure.

T. Schei (✉)
Finnmark University College, Alta, Norway
e-mail: torbjorn.schei@uit.no

K. B. Lindkvist, T. Trondsen (eds.), *Nordic-Iberian Cod Value Chains*,
MARE Publication Series 8, DOI 10.1007/978-3-319-16405-2_12

Because knowledge also in traditional resource-based industries is the most important production factor and learning is the most important process (Lundvall and Johnson 1994), we need a better understanding of factors furthering or impeding learning processes. The learning organization, an organization which facilitates learning and continuously transforms itself (Pedler et al. 1991), at the level of the firm (Penrose 1995) and the importance of networks at the level of industry, has been extensively researched in various academic traditions (Porter 1990; Clark et al. 2003; Malerba 2004; Fagerberg et al. 2006). Less has been done to identify knowledge and learning conventions as underlying drivers in social practises forming and modifying industrial action frameworks and long-term regional capacity building. By borrowing the action framework linked to the world of production concept from Storper (1997), it can be hypothesized that the matching and mismatching of knowledge and learning conventions are indicative of successful and less successful learning and knowledge creation.

The chapter proceeds with a brief description of EVT as a research topic. Then we describe how different approaches to theory of conventions can be combined in order to show how different conventions influence education infrastructure. We also discuss how these conventions are themselves modified and transformed from a combined interference of the political economy of education and industry and from everyday problem solving. A conceptual model for analysing an evolution of education infrastructure emerges. Here the matching of conventions among different groups of actors can explain the development of public infrastructure. The empirical and analytic parts of the paper follow the conceptual model. Firstly, the political economy of education is introduced. Then follows the political economy and practice of the fishing industry. Thereafter we discuss the subsequent development of education infrastructure in Northern Norway, including the parts and positions taken by government, academia and industry.

The analysis is built on secondary data available from the project complemented with primary data from personal interviews with former teachers in educational and vocational training (Fiskarfagskolen), industry, Directorate of Fisheries, historians at the Norwegian College of Fishery Science and sectorial NGOs.

12.2 Vocational Education in a Learning Region

Education and vocational training (EVT) is a research area which is interchangeably also referred to as Technical and Vocational Education Training (TVET). EVT serves the qualification of skilled workers for the employment system of an industry. The importance of EVT has increased with acceleration of technologies, economic change and dimensions of internationalization, and can be viewed as a decisive factor for innovation and prosperity in the international competition for quality (Rauner and Maclean 2008). EVT research is interdisciplinary using pedagogy, labour studies, sociology, economics and engineering to develop possible research questions, suggest methods and discuss results. Basic research has included development of

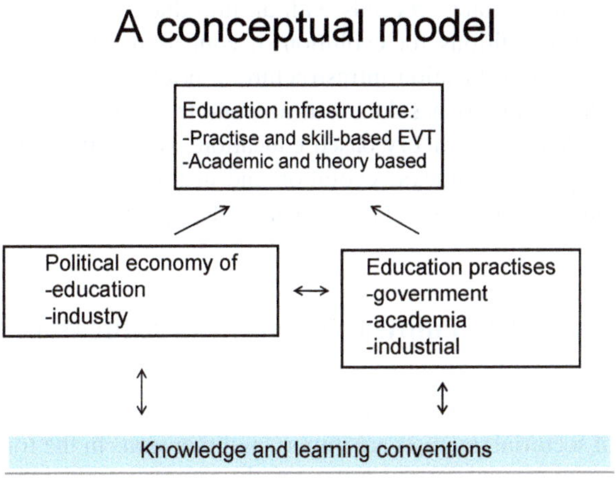

Fig. 12.1 A conceptual model of conventions in the formation of education infrastructure

EVT systems, plus design and evaluation of vocational training processes. Attention here is on the paradoxical erosion of a pioneer EVT institution in an era of national ambitions for education and research. The chapter focuses on how EVT appears to have 'slipped out of attention' in innovation policy and it studies the negative consequences not only for industry but also for the scope, variety and competitiveness of education in the market for human capital.

12.3 Conventions

An analysis of learning and knowledge conventions can be framed and structured along different traditions from Salais and Storper (1992) to the seminal work of Boltanski and Thèvenot as discussed in Chap. 2.

In this chapter the typology of the justification framework is applied to identify and structure knowledge and learning conventions. In the perspective taken by the literature on learning economies and territorial innovation, there should be a balance between practise-based and analytic (more academic) learning (Lundvall and Johnson 1994; Asheim and Coenen 2007; Gertler 2010). Furthermore in an institutional economic approach, conventions are not static but co-evolve with context and everyday problem solving. This is the agency-structure interdependence, where there is no such thing as a stable equilibrium but evolutionary processes driven by learning, habituation and creative destruction (Nelson and Winter 1982; Hodgson 2007; Martin 2010). In the analytical model (Fig. 12.1) there is an interaction and mutual constitution between the political economy of education and industry, education practises and conventions. This interaction forms an underlying mechanism of education infrastructure (lower part of Fig. 12.1). These dynamic processes in formation of the education infrastructure are materialized in the observable education infrastructure (upper part of Fig. 12.1).

The chapter proceeds with an introduction of the political economy and structural conditions for evolution of education infrastructure and the fish processing industry. Education infrastructure is briefly described, with an emphasis on EVT, where Norway has done a turnaround from pioneering EVT to a total dominance of academic and theory based education. Then a typology of knowledge conventions is applied to analyse evolution and justification of the current situation among the major stakeholders; government, industry and academia .

12.4 The Political Economy of Education

In a mixed, corporate economy like the Norwegian, the voicing and negotiating power of sectorial interests are important dimensions in the formation of public policy. In a series of publications following the research project "Education as nation building" financed by the Norwegian Research Council 1998–2002, the political dimension in the expansion and architecture of education infrastructure in the post-world-war II period is highlighted (Telhaug and Aasen 1999). Little or no reference is made to EVT or industrial needs. Focus is on primary education and conditions and discourses which made access to secondary schools a civil right in 1994 (and changed sectorial EVT into an ordinary secondary school and with fewer classes).

The political turn in the 1980s/1990s—concomitant with a shift from a centrally planned industrial development regime to endogenous growth perspectives (Piore and Sabel 1984)—also influences the education sector. Liberal and neoconservative forces make claims to education curricula and indicators of productivity are on the political agenda. This political turn is part of a broad discourse across the Western states about the future of the welfare state and its governance, and it marks an introduction of new public management perspectives and innovation policy which become important for both industry regulations and education.

12.5 The Political Economy of the Fish Processing Industry and Demand for Educated Labour

What Telhaug et al. (1999) characterize as the one-party state from 1940 to 1980s/1990s, in their description of education as a social democratic project, also comes with a deep influence on the Norwegian fisheries and fish industry. In 1937 a new law "The Raw Fish Act" (Råfiskloven) made fish price a negotiated entity, to be decided by representatives of The Norwegian Fishermen's Sales Organisation (Råfisklaget) and industry, but with exclusive right of the fishermen to take final decisions if the parties did not reach agreement on price (www.rafisklaget.no). The background is price fluctuations and social need and the aim is to stabilize prices and make the life of the fishermen more predictable (Hallenstvedt 1982; Christensen and Hallenstvedt 2005). After WWII fishermen organized themselves and took a strong corporate position (Finstad 2005) in a social democratic reconstruction and modernization. Much of the fish processing industry, on the other hand, became centrally planned. During the 1950s and 1960s this industry in the north (based on the rich cod fisheries) was 'modernized' and new freezing technology emphasizing frozen fillets was introduced (Vea 2009). Conventional dried and salted fish products, which historically had been the economic base for fisheries in the north (Richter-Hanssen, Chap. 3), were left in "the political shadow" (Finstad 2005). Furthermore, export licences were centralized to a few privileged organisations located at fishing centres on the West coast; Ålesund and Bergen, and the capital area in the south-east. In this situation most of the industry produced for centrally planned sales organizations with price paid per unit, not according to quality (Finstad 2005), which cut off the direct market information network that might have inspired the processors to develop products and marketing strategies according to consumer demand (Trondsen Chap. 9). This publicly regulated, Fordistic industrial practise provided few incitements for educated labour. Furthermore, profitability failed. During the 1970s and 1980s state subsidies increased and reached a dazzling 4 billion NOK in the late 1980s. The loss of profitability in an industry built on one of the richest fish resources in the world can be traced to a strategy for standardised products exposed to international price competition with few concerns for variations and stochasm in availability of the resource or for variations in qualities. Low margins were met with scale economic investments undermining the carrying capacity of the resource (Finstad 2005; Trondsen, Chap. 13).

In 1991 export restrictions were eased and private enterprises were granted the right to export. However, a series of resource particularities and structural boundaries still make the fish industry different. Industrial profitability is on average still low, providing few opportunities for paying a wage premium for educated labour and subsequent low incitements for education.

12.6 Education Practises—Evolution of Education Infrastructure

Compared to agriculture the education infrastructures for the fishing industry came late and the level of education in the industry is significantly lower than in agriculture (Normann and Christensen 1988; Berg 2006). However, the fishing industry is founded on a culture of hunting and harvesting an open access resource, not sowing and cultivating private land (Handegård 2010). Major elements in the evolution of regional and sectorial education structure in North Norway, Finnmark and Troms counties, are illustrated in Table 12.1.

The time-line in Table 12.1 is categorized into three periods: before WWII when LIF (Statens Forsøks- og lærebruk) was established; 1970–2000 which are years of expansion in higher education; and 2001–2010 representing the last 10 years. In order to understand the background for a skewed development of education infrastructure,

Table 12.1 Institutions in education infrastructures in the North; Finnmark and Troms counties

	Before WWII	1970–2000	2001–2010
Higher education		B C	F
Secondary school		D E	
EVT	A		
On job training			G

A = 'Statens Forsøks- og Lærebruk'—LIF, sectoral EVT institution established in 1933
B = The Norwegian College of Fishery Science was founded in 1972 and located in Tromsø, as a faculty in the University of Tromsø (founded in 1969)
C = The University College of Finnmark founded in 1973
D =Secondary school became a civil right in 1994 by the 'Kunnskapsløftet' reform
E = Statens fagskole for fiskeindustri (a modified version of LIF) became an 'ordinary' secondary school; VardøVideregående skole, in 1994
F = The Salted Fish Academy project, a field course at the bachelor level, initiated in 2010
G = Enterprises in the regional salted fish industry state that they must be self-supporting in EVT. (Interviews 2010)

explanations are sought in the practises and conventions of the major stakeholders: government, industry and academy.

The county of Finnmark is the northernmost and the thinnest populated county in Norway, located up till 71° north, bordering Sweden, Finland and Russia. Regional trade houses and industry, working across borders with Russia, expanded until the Russian revolution in 1917 when the eastern trading route and the trade with Russian Pomors were closed. Efforts to develop alternative trade routes to the West met competition from well-established west-coast and European trade houses. From being a geographically favoured area at that time Finnmark became peripheral and disadvantaged (Richter-Hanssen, Chap. 3). The financial breakdown in the 1930s added problems to an economy already in crisis. Social conditions were alarming and in a letter from the Directorate of Fisheries in 1933 the question was raised whether something fundamental could be done to improve the foundations of economic development. This sparked the establishment of a national experimental station (forsøksbruk) and vocational education school (lærebruk) (LIF) Vardø close to the Russian border. Vardø was a vital fishing community in Finnmark at the time, built on a combination of rich fishing resources and the Pomor trade.

12.6.1 The National Experiment and Vocational Education School 'Statens Forsøks- og Lærebruk' (LIF)

The National Experiment and Vocational Education School 'Statens Forsøks- og Lærebruk' (LIF) later also named 'Statens fagskole for fiskeindustri' was initiated in 1933 as a pioneer in national and international education for the fishing industry.

As a token of a new era the school was established in the localities of a former trade house and leading fishing enterprise. By 1938 the school was a success. Students were reported to gain leading positions in industry and administration and "it is told that in these enterprises products are outstanding" (Norman and Christensen 2005, p. 58). It was also referred to students who saw the opportunities in new products and technologies and started new enterprises. In late 1944 buildings and infrastructure in most of Finnmark were burned and destroyed. After WWII, a centrally planned rebuilding of Finnmark took place and a 'modernization' and large scale industrialization of the fishing industry followed in the 1950s and the 1960s (Finstad 2005). LIF, however, never got the financial means to regain a national position. In 1962—at the heights of scale-economic, standardised production philosophy- the school lost its status as an experimental and development institution, and its mandate was narrowed to EVT only.

In 1994 a national reform removed the last part of industrial identity by turning what was then named 'the vocational school' (Fagskolen) into a standardized 3 year secondary school 'Vardø videregående skole'. Seafood became one of several topics open for specialization in the second and third year, but with a limited curriculum and a reduction of class hours. Students enrolled in the vocational school at the time of the reform protested, pointed to the value of the unique brand– 'The vocational school' (Fagskolen)—and pleaded to have their diplomas still titled this way. The application was granted for those already enrolled. In the following year, 1995, applications for the new secondary school dropped. It had lost its appeal to those employed or who were on the brink of getting employed in the industry and seeking formal vocational education: "These potential students were not interested in a generalized curriculum. For them the topics of dairy and general food processing were perceived as an unnecessary delay". (Interview with former teacher at the vocational school 2010). The reform in 1994 came with a more generalized and theoretical curriculum, which has been criticized and related to increasing numbers of drop-outs in secondary school (Gjerstad 2010).

12.6.2 Government

At the level of government Gammelsæther (2002) identifies a confounding mix of interests in forming a national education and research policy. While the Ministry of Education and Research is 'the owner' of the sector, the Ministry of Industry and Trade and the Ministry of Regional Affairs have exercised strong interest in the sector when it comes to its position in regional development. The Department of Fisheries and its Directorate of Fisheries have, on the other hand, lost position in educational affairs. While the Norwegian School of Agriculture was well situated under the Ministry of Agriculture, the part played by the Ministry of Fisheries in relation to education infrastructure has been less clear (Normann and Christensen 1988; Handegård 2010). In analysing the funding problems faced by LIF in rebuilding after war damage and in necessary upgrading of teaching facilities, Norman and

Christensen (1988) point to unclear and shifting parts and responsibilities in the government to solve this problem. While EVT was initiated and planned by government sectorial administration, under the Directorate of Fisheries in Bergen, the responsibility for the school was later transferred to Oslo and the Ministry of Education. Here the school lost its foundation in the industry and became an integrated part of national reforms in secondary school, focusing the welfare aspect of and access to education in general (Telhaug and Aasen 1999). The history of the institution (Normann and Christensen 1988) indicates that the factual situation was not well known in government outside the Directorate of Fisheries (located in Bergen). It appears that, when the responsibility for sectorial EVT was transferred to the Ministry for Education in Oslo, it added an industry-government divide.

12.6.3 Higher Education

In 1972 the Norwegian College of Fishery Science was established as part of the University in Tromsø. Developing a national education and research institution for supporting the industry and sectorial public policy became however difficult, according to Handegård (2010) who describes two major challenges. These are briefly described as an illustration of a perspective where political intentions of establishing a national coherent centre of competence were not realized. First, the number of staff in Tromsø never passed 50 (including technical support, excluding scholarships). In reflecting over the capacity of the new institution to serve industrial demand for education, research and support public policy formation, Handegård (2010) makes a comparison with the Norwegian Agricultural School with its staff of 320. A second challenge became what Berge and Gammelsæther (2002) denote as academic drift, where disciplinary and academic oriented focuses dominate the priority of resources. The phenomenon of academic drift is exemplified by the recruited economists. While recruited to a national institution for fisheries and fish industry there was a gradual drift in teaching and research towards general economic topics (Handegård 2010:240).

In addition to the Norwegian College of Fishery Science a national system of university colleges was developed during the 1970s and 1980s to the extent that it was found remarkable in a European context (Gammelsæther 2002). At the Finnmark University College several efforts were made to develop studies for the fish industry and marine sector. Among these were Fishing Industry Management (30 ects) 1982–1984 and a Bachelor in Aquaculture from 1990 to 2005, both, however, terminated due to lack of applicants. "There has been an interest in establishing education directed toward the largest industry in the country, but it has proved difficult. The industry in 1980 was composed of smaller enterprises with little formal administration that did not seek education" (Berg 2006, p. 98).

12.6.4 Industry

The part of industry in developing education infrastructure has been moderate. The history of the LIF illustrates that the voicing for necessary funding and complaints about a deteriorating physical and technical standard in relation to industrial needs, was mainly driven by the board of the school, composed of sectorial administration, regional politicians and representatives for the industry—the industry as such was less active. Particular for LIF is its foundation in the physical infrastructure of a former fish plant, and the combination of industrial and pedagogical activity. This appears to be an ideal combination of learning by best practise and skill building. However, the need for a regular supply of fish became increasingly problematic when the imbalance between fish resources and industrial capacity became a limiting factor in industrial profitability, because neighbouring processing plants identified the school as a rival in the acquisition of scarce input. This became problematic for the teaching which depended on predictable access to a variety of fish species. After a formal complaint it was decided that the school could not pay a price premium and establish long-term contracts with the fishermen (Normann and Christensen 1988). As pointed out by Finstad (personal interview in 2010) the conventional part of the processing industry was not well organized in the years after WWII. Furthermore, enterprises were small and many were struggling to survive. They went out of business following the imbalance in industry capacity due to rapid expansion in government supported large scale frozen filets plants. After the closing of LIF in 1994 the physical infrastructure was left under-utilized. In 2008 a private company proposed to take over the unused localities in the school in order to combine industry and education endeavours. This initiative was, however, turned down by county authorities with responsibility of secondary education (personal interview with leader of this firm in 2010).

12.7 Conventions and Justification in Education Infrastructure

An approach to structuring and categorizing different norms and justifications in education is given by Boltanski and Thévenot (2006) and their analytic framework and typology. Table 12.2 is an application of this framework to education infrastructure. The table categorizes conventions among stakeholders in the triangulation of government, industry and academia, and illustrates differences and similarities. With respect to similarities, Storper (1997) has identified the matching of conventions among actors as a key dimension in competitive action frameworks; which for the case of education infrastructure can be translated to dominating priorities in the governance of education.

For some of the categories of justification an application to knowledge conventions appear straight forward: e.g. the world of inspiration, which in a justification of knowledge conventions can be applied to defend free research (and the potential for value creation by inspired professional creativity). The domestic world points to justifications defined within the hierarchies in different production systems and

Table 12.2 Knowledge conventions in the typology after Boltanski and Thévenot (2006) across government, industry and academia

Knowledge conventions	Government	Industry (fishermen organizations, processing and export organizations)	Academia
Inspiration			*Free research*
Domestic		*Skill-based production (sub)systems*	*Research paradigms and disciplinary hegemonies*
Renown/opinion/ public	*International publication*		*International publication*
Civic	*Education as a welfare good and regional development tool*		*Education as (beneficial) transmission of theory (to the less knowledgeable)*
Market	*Generic marketing*	*Generic marketing*	*Specialized applicable (tradable) research*
Industrial	*Health and resource considerations*	*Skill based, internal training adapted to volume oriented production*	*Specialized (efficiency enhancing) research*

spheres: in the words of Boltanski and Thévenot "... one must, to evaluate someone's worth, know his place in the network of dependencies from which this person draws his own authority", (2006, p. 370). In knowledge production we find localised hierarchical systems in the skill-based production systems, sometimes in the form of a guild. In academia professional authority is defined by research paradigms and disciplinary hegemonies: e.g. the one between orthodox and heterodox economics in which domestication is present to the extent that a lack of cross-references between complementing fields of research is considered a problem (Blaug 1992; Rutherford 1994).

The civic world is the world of the common good. Table 12.2 illustrates a situation where government provides infrastructure for education which is seen as welfare good. Academia provides knowledge to the less knowledgeable. Also, in the world of the renown, conventions in government and academia match in the quest for (and priority of) international publication. The market world is one of coordination, buying and selling through the mediation of scarce goods. Here academia has knowledge to sell in the form of specialized research (or tailor made courses). When knowledge needs are acknowledged, industry or government is characterised by willingness to pay for research. There is mutual trust between the parties.

The situation is one of matching conventions and complementing interests. This situation is well documented in the cases of industrial clusters and emerging and profitable high-tech industries (Porter 1990; Asheim and Coenen 2007; Gertler 2010).

Such a matching is also found in the salted fish industry, but as documented by Lindkvist (2010), mistrust of research institutions and domestic knowledge conventions appear as a hindrance in the salted fish innovation system. However, an unexpected discovery was made in common preferences for generic marketing among the Norwegian parties. Due to the Raw Fish Act fishermen organizations have a strong position in the industry, and considerable financial means. There is also a historical strong link between fishermen organizations and government (Finstad 2005). This is one explanation why the governmental development bank and Innovation Norway offers resources dedicated to marketing Norwegian fish products. The result of these matching conventions are seen in e.g. current campaigns for Norwegian salted fish in the Spanish market, as described by Hernández (this book, Chap. 4).The industrial world is the world of efficiency, "measured on a scale of professional capabilities" (Boltanski and Thevenot 2006, p. 372). To describe the objects of the industrial world a productivity guide can be applied, and outstanding persons are the experts. In the government sector, health and resource control are priorities. Historically EVT has also been given a high priority, illustrated by the founding of a pioneer EVT institution by the Directorate of Fisheries –this no longer exists.

Contrary to studies of high-tech (successful) industries a coherent set of knowledge conventions between government, industry and academia is not found with respect to education infrastructure. To paraphrase Storper (1997); without a coherent set of knowledge and learning conventions there can be no well-coordinated action framework, which support the findings of Lindkvist (2010) describing the innovation system as imaginative. Notwithstanding a situation of fragmented conventions, some partial-coherence can be found, first and foremost in matching conventions between government and academia, explaining the development of academic oriented education infrastructure.

12.8 Conclusions

We have found that the erosion of EVT is due to an insufficient funding, in spite of reports describing the performance of former students as outstanding. In the same period, research-based education infrastructure has expanded and taken an increasingly important position in regional and innovation policy. The two northernmost counties in Norway; Finnmark and Troms, with about 230,000 inhabitants, are on a national level in infrastructure for higher education and research institutions. In addition, the Norwegian College of Fishery Science is located in the region.

However, the salted fish industry has been and still is composed of relatively small enterprises, characterised by a low level of formal competence and no tradition for seeking higher education. The dominating cost-oriented industrial convention did not demand so many specialized higher educated people in market-oriented production. An increasing mismatch of knowledge conventions emerged also between the fishing and processing industries and the academic oriented development of education infrastructure. Matching knowledge conventions are only found across government and academy on the justification of education infrastructure as a welfare good per se for students, and in focusing on international publication for teachers. These are examples of coherent sets of conventions defining action framework in education infrastructure evolution, however at the cost of an increasing academy-industry divide.

If this sectorial case study reflects national education and innovation policy in general, showing a mismatch in learning and knowledge conventions between industry and a coalition of government and academy, then it is an indication of structural weaknesses in the national innovation system. In addition, the important contribution of upper secondary school in industrial competence appears down-played in the innovation system literature, which is a challenge for the research community.

References

Asheim B, Coenen L (2007) Face-to-face, buzz, and knowledge bases: sociospatial implications for learning, innovation, and innovation policy. Environ Plan C Gov Policy 25:665–670
Barney JB (1996) Gaining and sustaining competitive advantage. Addison-Wesley, New York
Berg K (2006) Fremmed fugl? Historien om Høgskolen i Finnmark 1973–2006 Høgskolen i Finnmark, Finnmark
Berge DM, Gammelsæther H (2002) Høgskolene i innovasjonspolitikken. In: Gammelsæther H (ed) Høgskoler til besvær. Når statlige reformer møter lokale institusjoner og ambisjoner. Fagbok forlaget, Bergen
Blaug M (1992) The methodology of economics: or how economists explain, Cambridge surveys of economic literature. Cambridge University Press, Cambridge
Boltanski L, Thèvenot (2006) On justification: economies of worth. Princeton University Press. Princeton. (French edition 1991: De la justification. Les économies de la grandeur)
Christensen P, Hallenstvedt A (2005) I kamp om havets ressurser. Norges Fiskarlags historie. Norges Fiskarlag, Trondheim
Clark G, Feldman M, Gertler M (eds) (2003) The Oxford handbook of economic geography. Oxford University Press, New York
Dewey J (1925) Experience and nature. Open court, 2012 edn. N.Y Courier Dover, New York
Fagerberg J, Mowery D, Nelson R (eds) (2006). The Oxford Handbook of innovation. Oxford University Press, Oxford
Finstad BP (2005) Finotro. Statseid fiskeindustri i Finnmark og Nord-Troms -fra plan til avvikling. dr. art Universitetet i Tromsø, Norway
Gammelsæther H (ed) (2002) Høgskoler til besvær. Når statlige reformer møter lokale institusjoner og ambisjoner. Fagbokforlaget, Bergen
Gertler MS (2010) Life sciences and regional innovation: one path or many? Eur Plan Stud 17(2)
Hallenstvedt A (1982) Med lov og organisasjon. Organisering av interesser og markeder i norsk fiskerinæring. Universitets forlaget, Oslo
Handegård O (2010) NFH, FTFI og NFFR Lokaliseringsbeslutningene 1945–1975. Noen organisasjonsmessige endringsprosesser 1976–2009. Norges Fiskerihøyskole, Tromsø
Hodgson GM (ed) (2007) The evolution of economic institutions. A Critical Reader. Edward Elgar
Lindkvist KB (2010) Mistrust and lack of market innovation. Eur Urb Reg Stud 17:31–43
Lundvall BÅ, Johnson B (1994) The learning economy. J Ind Studies 1:23–42
Malebra F (2004) Sectoral systems of innovation: concepts, issues and analyses of six major sectors in Europe. Cambridge University Press, Cambridge
Martin R (2010) Rethink regional path dependence: beyond lock-in to evolution. Econ Geogr 86:1–27

Nelson RR, Winter SG (1982) An evolutionary theory of economic change. Harvard University Press, Harvard
Normann H, Christensen P (1988) "Lærebruket". Statens Fagskole for Fiskeindustri 1938–1988. Norges Fiskerihøyskole, Universitetet i Tromsø
Penrose E (1995) The theory of the growth of the firm. Oxford University Press, Oxford
Piore M, Sabel C (1984) The second industrial divide. Basic Books, New York
Porter ME (1990) The competitive advantage of nations. Macmillan, London
Rauner F, Maclean R (2008) Handbook of technical and vocational education and training research. Springer, Germany
Rutherford M (1994) Institutions in economics. The old and the new institutionalism. Historical perspectives on modern economics, Cambridge University Press, Cambridge
Storper M (1997) The regional world: territorial development in a global economy. Guiford
Telhaug A, Oftedal A, Aasen P (1999) Introduksjon. Mot et nytt kunnskapsregime? In: Telhaug A, Oftedal A, Aasen P (eds) BÅDE-OG 90-tallets utdanningsreformer i historisk perspektiv. Cappelen Akademiske forlag, Oslo

Chapter 13
Value Chains' Constraining Effect on Industrial Conventions and Market Adaptions

Torbjørn Trondsen

Abstract Value chains constrain the business transactions' capacity to maximize market value from limited resources. Seafood markets exhibit a strong demand for fresh, high-quality seafood, yet empirical studies show large variations in value chains' adaptation and exploitation of market opportunities.

This paper shows that both the distance in space and time between the wild harvest and the consumption regions and the choice of preservation methods are important structural constraints for market adaptation. A comparative study of the Norwegian and Icelandic exports also shows that the value chains' conventions, geographically rooted in harvesting and processing regions upstream, have a significant effect in shaping differences in the value chains' market adaptation downstream.

Keywords Constraining effects · Conventions · Market adaptions · Fish exports · Cod value chains · Environmental adaptation · Governance · Iceland · Norway

13.1 Introduction

Business activities that cultivate advantages by satisfying the needs of customers more efficiently and effectively than competitors have been identified as a major driver of superior business performance. Rapidly changing environments require quick and continual adaptation by individual firms (Cano et al. 2004). The individual firms' capacity to adapt to consumer demand is however constrained by the embedded value chains which are connecting regions of raw material sources and production to consumption (Grunert et al. 2010). The configurations of value chains shape an atmosphere that encourages transactions linking suppliers to buyers. Markets dominated by production-oriented controlled value chains are for example related to mismatches between supplied and demanded product qualities (Fisher 1997; Lindkvist et al. 2008; Lindkvist and Sánchez 2008).

T. Trondsen (✉)
Norwegian College of Fishery Science, University of Tromsø, Tromsø, Norway

Department of Geography, University of Bergen, Bergen, Norway
e-mail: torbjorn.trondsen@uit.no

K. B. Lindkvist, T. Trondsen (eds.), *Nordic-Iberian Cod Value Chains*,
MARE Publication Series 8, DOI 10.1007/978-3-319-16405-2_13

These mismatchs may rely on the constraining characteristics of the wild fish harvesting industry in general, but also from different geographics of the regions where the value chains are embedded. This paper will explore the significance of these two sets of constraining factors which empirically includes a comparative case study of the Norwegian and Icelandic export of cod products to the European market. This way of analysing a value chain as integrated with its regional context is in line with Crevoisier (2004) and Gereffi et al. (2005) who show that the territorial maneuvering room for entrepreneurship is shaping innovative processes and codetermines their evolution. Crevoisier (2004) considers this regional function as a matrix for economic development. Territories are socio-economic arenas (values, networks, institutions and their material production structures) for decision making. This underpins the value chain activities related to production, distribution, consumption and regulation of the seafood value chain. Wild seafood resources are spatial or geographically located in most countries by law to the commons, managed as a public trust by the state on behalf of the people (Stevens 1980–1981, UN Convention on the Law of the Sea[1]). Harvest regulations as a basis for formal legal arrangements compel producers to follow rules and practices which again are spatially situated. This may be considered by authorities as conducive to realization of national values and goals in the sector. Other formal organizations, private/public or profit/non-profit, located in between producers and consumers, also play specialized roles related to value chain transactions. All such rules, whether public and direct, or silent or indirect, are influenced by regional customs.

Quite often value chains are used merely as a descriptive device to show 'links' between activities. This paper adds to the existing literature by exploring the role of different kinds of structural arrangements in the value chains that are constraining market adaptation in time and space. Such constraining of geographically rooted factors are described as fundamental mechanisms of a market economy (Crevoisier 2004; Andresen 2009). More precisely the paper shows how value chain structures of wild seafood trade are constraining market adaptation in central consumption areas in a seafood industry relying on wild natural resources harvested in remote North Atlantic areas.

As a food source, seafood is increasingly highly valued and is recognized as having great nutritional health benefits, especially among more highly educated consumer groups (Trondsen et al. 2004a; Sánchez 2011). These consumer groups are globally characterized as emphasizing convenience, food quality, and flavour (Myrland et al. 2000; Trondsen et al. 2003b; Scholderer and Trondsen 2008; Sánchez 2011). For example, whiteness and fillet thickness, together with prepared product forms convenient for cooking such as light salted or fresh fish, are considered particularly attractive product qualities of North Atlantic cod and are highly valued in the Spanish market, especially among wealthy regions such as Catalonia (Espinosa Segui,

[1] Article 1 "The coastal State shall determine the allowable catch of the living resources in its exclusive economic zone" "to ensure. the living resources ... is not endangered by over-exploitation..., as qualified by relevant environmental and economic factors, including the economic need of coastal fishing communities."

Chap. 5; Lindkvist et al. 2008). Such changes in consumer attitudes have challenged seafood suppliers to adapt their offerings to these new market preferences. As the success of Icelandic firms compared with Norwegian firms in the Spanish codfish market illustrates, the most market-adaptive firms in an environment achieve competitive advantage in the marketplace (Lindkvist Chap. 7; Lindkvist et al. 2008; Trondsen Chap. 9). The seafood industry's dependency on natural resource harvesting, heavily influenced by ecological fluctuations driven by natural variation, harvesting effort, and government regulation, is described as the most important barrier to such market orientation (Trondsen and Johnston 1998). Value chain management in such unstable environments can be likened to manoeuvring ships through rough seas—these conditions constrain the opportunity for value chain specialization according to consumer preferences and demand.

Management strategies are in this paper explored in the context of a Structure–Convention–Performance (SCP) value chain model in which the term 'convention' has replaced the term 'conduct' in the standard SCP model (see Chap. 2). A business convention is defined here as an accepted way of thinking, behaving, or doing business in particular business groups, in contrast to conduct that is limited to actual organizing and doing business. The study draws on insights from the present author's 40 years of study and experience of the fishing industry, as well as insights from fishery science, institutional economics, economic geography, business, and marketing.

The following sections of this paper apply the value chain model (see Fig. 2.1, Chap. 2) in a comparative analysis of business conventions and performances in the Norwegian and Icelandic cod sector.

13.2 A Contemporary Comparative Analysis of Norwegian and Icelandic Fish Exports

Freshness is the most highly valued fish quality in all consumer markets. In 2008, the Norwegian export share of fresh salmon (whole and fillets) was about 86 % of the total salmon export while the comparable export share of fresh wild cod was only 13 %. Most wild cod is exported as lower-value products preserved in salted, dried, and frozen forms. An important difference in market adaptation between these two value chains is obviously that the aqua cultured salmon industry can continuously satisfy a demand for fresh fish by controlling slaughtering according to incoming market orders, whereas the cod industry is more dependent on harvesting a wild, fluctuating, natural resource.

Another variation in market adaptation exists between the value chains supplying the same kind of Atlantic cod from Norway and Iceland, thus indicating that other territorial factors also are constraining market adaptation.

Table 13.1 shows Iceland's and Norway's export profiles for cod products. In both countries, fresh fillet products are the highest valued utilization of the cod harvest calculated in unit values live fish equivalents. Fresh fillets are exported ice chilled

Table 13.1 Average exports of cod products. (from Iceland and Norway calculated in live weight and 2008 €)

	1990–1999				2000–2008			
Iceland's cod exports	Av. million kg	Share (%)	Av. € /kg	Rel. Value (%)	Av. million kg	Share (%)	Av. € /kg	Rel. Value (%)
Fresh fillets	10	4	1.28	100	26	10	2.51	100
Consumer packs					13	5	2.31	92
Sea frozen fillet blocks	6	2	1.33	104	2	1	2.17	86
Salted split	80	28	1.55	121	65	26	2.13	85
Sea frozen fillets	36	13	1.33	104	34	13	1.95	78
Salted fillet	18	6	1.37	107	22	9	1.91	76
Fresh unprocessed	13	5	1.36	106	15	6	1.81	72
Land frozen fillets	72	25	1.26	98	60	24	1.57	63
Land frozen fillet block	43	15	1.01	79	10	4	1.27	51
Clipfish	3	1	1.28	100	5	2	0.62	25
Total	282	100	1.31	102	251	100	1.96	78
Norway's cod exports								
Fresh fillets	4	1	1.07	100	9	3	2.82	100
Stock fish	27	6	1.70	159	25	7	2.48	88
Fresh unprocessed	25	5	1.14	107	29	8	2.14	76
Clipfish	116	26	1.39	130	113	33	1.90	67
Salted split	112	25	1.21	113	62	18	1.85	66

Table 13.1 (continued)

	1990–1999				2000–2008			
	Av. million kg	Share (%)	Av. € /kg	Rel. Value (%)	Av. million kg	Share (%)	Av. € /kg	Rel. Value (%)
Frozen fillets	119	26	1.10	103	49	14	1.83	65
Salted fillets	11	3	1.11	104	4	1	1.82	65
Frozen unprocessed	20	4	1.08	101	27	8	1.75	62
Frozen fillet blocks	63	14	1.35	126	26	8	1.44	51
Total	453	100	1.27	119	345	100	1.94	69

Conversion factors are from the Norwegian Directorate of Fisheries and export data from Seafood Norway and Icelandic Statistics both as higher-priced cuts of loins and tails or as whole fillets that include the thin belly. Iceland responded to the demand for fresh fish by increasing the fresh fillet share of its total cod export quantities from 4 % in the 1990s to 10 % in the 2000s; the corresponding shares for Norway were 1 % and 3 % (all calculated in live weight). The Icelandic industry achieved this growth despite its logistic comparative disadvantage as an island dependent on more expensive air transport to move fresh fish with a limited shelf life to the main European markets, compared with Norway's advantage of cheaper road transport to the same markets. Iceland also has a disadvantage in the wet climate for production of the valuable dried stockfish, highly prized in the Italian market, whereas Norway, in the Lofoten area, has an absolute climate advantage for drying in an ideal temperature. Despite these disadvantages, however, Iceland gained a higher average value per kilogram for its total export of cod products in both the 1990s and the 2000s, especially enjoying price advantages in the market for higher-priced salted split cod, sea frozen (fresh frozen) fillets, and salted fillets. The Norwegian cod value chain has a greater presence in lower-priced salted split and frozen and salted fillets. The Norwegian fish industry has however gained a dominant export share of clipfish thanks to a controlling position in the cod value chains supplying the Brazilian and Portuguese markets (Trondsen 2012).

The differences between the export profiles of Norway and Iceland for the same fish species indicate territorial differences in constraints and business conventions between the two value chains supplying many of the same markets. This divergence is explored in the following.

13.3 Characteristics of Value Chains for Wild Fish

Seafood can be characterized according to many heterogeneous product qualities related to species, size, freshness, and the use of preservation methods. The commercialization of seafood products relies on heterogeneous consumer market preferences matching the traded product qualities. In practice, therefore, the seafood trade is highly specialized in sorting highly heterogeneous raw materials into the appropriate homogeneous product qualities to match homogeneous demand from specific consumer groups (Bestor 2004; Grunert et al. 2010).

The process of market adaptation through homogenization and standardization of products from heterogeneous raw materials along seafood value chains requires coordination of numerous production and distribution stages, during which raw fish and other input factors are transformed into products for which consumers are willing to pay competitive prices.

Wild fish are harvested by fishing vessels, slaughtered, sorted according to species, size, and freshness, cut into consumer-ready sizes, packed, and distributed to final consumers in an unstable embedded environment. Increasing distance and distribution time between the harvesting and consumption areas create a need to preserve product quality by methods such as chilling, salting, drying, freezing, or canning. These operational value-adding steps are organized in specialized production units in value chains that form strategic groups of similar operations, with companies having different degrees of vertical integration (Porter 1990). Seven main strategic groups that follow the same main bundle of conventions can be identified in fish value chains, which can be further divided into specialized subgroups: (1) fishing vessels (harvesting, icing or freezing); (2) fishing and processing vessels (harvesting, filleting, freezing); (3) first-stage processing, e.g., harbour auction (sorting, icing, packaging, distribution); (4) second- stage processing (filleting/cutting, salting, drying, freezing, distribution); (5) third- stage processing (drying salted fish (clipfish), canning, distribution); (6) fourth- stage processing (consumer packaging, wholesale distribution); (7) final-step processing (distribution to retail, hotels, restaurants and catering firms)

13.4 Market Adaptation of the Cod Value Chain from Norway and Iceland

The Norwegian cod value chain has over time been concerned more with specialized cost efficiency in catch and production than with market differentiation (Trondsen 1985; Lindkvist et al. 2008). In particular, the first operational stage in the Norwegian seafood value chain specializes in pushing traditional semi-processed cost-efficient products into the market rather than in offering consumer markets the ready-to-cook products and services that are desired.

The industrial conventions dominating the specialized Norwegian value chain differ from those of Iceland, where much more specialized consumer market-oriented production and value adding are practiced (Trondsen Chap. 9). This does not mean that Norway lacks competitive strength in cod fish export markets, which include about 150 countries. Rather, this means that Norway controls a fish resource highly valued in international markets that is efficiently exported as semi-processed raw materials for further processing in homes and factories closer to the final consumer. The Norwegian industry's industrial and commercial conventions are therefore dominated by relationships and transactions in value chains that move raw materials toward further processing links closer to the consumer markets. The lower commercial consumer market orientation in the Norwegian supply chains means less focus and value adding in servicing the end consumers and their suppliers of consumer-ready products and qualities.

This lack of adaptation to specialized consumer market conventions by the Norwegian cod fish processors gives processors and distributors further down the value chain a better window of opportunity for value adding: The European importers of Norwegian semi-processed products can add more value by processing consumer-ready products and services according to market conventions than from the more preprocessed products from Iceland (Trondsen Chap. 9). Many Norwegian fishery firms have tried to become more market oriented, but without success because, despite their willingness, the value chain structure makes this change difficult. Therefore, explaining the market performance of Norway's seafood value chain requires examining the various interlinked conventions embedded in the chain. To provide context for these conventions, I first look at key environmental imperatives that govern fishing activities.

13.5 Environmental Imperatives

The first important constraint on transaction control in fish marketing is the limited shelf life of high-quality freshly caught fish. In terms of shelf life of good-quality fresh produce, fish is more like milk than animal meat, in that it can be preserved on ice for only about 10–14 days. Marketing of the most highly valued fresh fish is therefore limited to a geographic transportation distance of 3–7 days from the point of harvest and slaughtering to the retailer, who may require a 7-day sales period. Therefore, although fresh fish can be distributed by road from Norway to many parts of Europe, the available timeframe requires expensive airfreight for the same fish to be transported to Asia or North America. The fresh fish marketing chain faces tough competitive pressure to deliver the fish to the final customer in good condition, according to consumer consumption. Entry barriers into the fresh fish value chains are low, because the industry structure consists of numerous small-scale fishing vessels and fish auctions. The competitive pressure on seafood trade transactions generated by the low entry barriers between the traders enforces intense rivalry in their marketplaces.

Another important ecological constraint is the strong pressure on the transaction pattern caused by seasonal fish landings, which is greatest when the catching costs are lowest -during the spawning or feeding seasons (mainly in winter and spring) when the fish are concentrated in big shoals. Counterparts to this ecologically driven fish-landing pattern are the ecological conventions dominating coastal fisheries management. Fishers can secure satisfactory profits by bundling together different harvesting seasons over the year, unrestricted by a need to carry out the next processing stage in the value chain and thus violation the conventions further down the value chains. In addition, the combination of harvesting seasons may vary from year to year, depending on the ecological balance between fish species and catch availability. Variation in supply therefore forms a major structural constraint on the fish marketing conventions that control transactions downstream in the value chain. Seasonal fluctuations in wild fish availability vary across regions and are an important factor underpinning value chain comparative advantage and configuration.

13.6 Value Chain Environmental Adaptation

The great distances between the main consumer markets in major cities and population centers in the US, Europe and Asia and the major fish resources in remote areas in the northern and southern hemispheres form a major structural constraint on the fish value chain configuration and convention development. Harvesting of wild fish, especially by deep-seas trawlers, often takes place at a distance of several days sailing from their ports, which limits technological possibilities for offering fresh fish on consumer markets. Specialized industrial conventions have been developed to preserve fish quality over time, such as freezing, salting, canning, and drying as storage buffers, to match remote and seasonal harvesting periods to the more stable demand in consumer markets. For example, such production conventions focusing on preservation had already been developed by the Basque in the Grand Banks cod fisheries off eastern North America which, according to Kurlansky (1999, p. 22), took place even before Columbus' arrival in 1492; they salted the cod in the fishing vessels using salt brought from Spain. Salt and salted cod later became important commodities in Norwegian–Spanish trade, with Norway importing salt from Italy and then from Spain to salt the cod that was then exported to Spain (Richter Hansen Chap. 3). Supplying preserved fish to markets requires that consumers or their fishmongers develop industrial conventions focusing on thawing, desalting, or dewatering the preserved cod to satisfy consumer conventions for ready-to-eat form. For many recipes, this process also requires mixing the fish with local raw materials such as olive oil and tomatoes .

By contrast, coastal fisheries, where harvesting takes place at only a day's sailing distance from land, have developed conventions for supplying fresh fish markets. Most of the catch landed by coastal fleets from EU countries and Iceland are sold fresh chilled, taking advantage of the premium prices for fresh fish. The fresh fish trade is supported by fresh fish auctions, where price-quality communication between customers and fishers reflects matching conventions of actual market opportunities and competitive market pressure. Auction mechanisms provide fishers with the most competitive commercial conventions through market communication and transaction experiences (Trondsen et al. 2003a; Trondsen and Young 2006).

Norway, where the coastal fleet landings are channelled into the fresh fish supply chain only to a limited degree, is an exception to the European pattern. Therefore, other structural territorial barriers must be operating.

13.7 Conventions of Seafood Value Chains

According to standard marketing theory, the competitive market-oriented conventions controlling high performance in fish value chains in order to create fresh, high-quality products may be based on specialized differentiation of rare combinations of product qualities and services demanded in the most attractive growing markets. The commercial conventions may be differentiated and protected by bundling marketing opportunities for high-quality and rare product characteristics, such as attractive product freshness and forms, species mix, supply stability, service, distribution, promotional support, and financing and price. New technologies also offer opportunities to extend the shelf life of fresh fish, by means such as super chilling (down to − 4 °C), new freezing and thawing methods, modified atmosphere and vacuum packaging.

Organizations that engage in competitive consumer-oriented marketing have to control the use of matching and skilled input resources (capital and human resources) through a strong innovative business orientation that is supported by technology providers and strategic investments in distinctive capabilities (Day 1994; Narver et al. 2004). This means that the focus of the need for industrial conventions has to be more market related. Key success factors in marketing fresh, high-quality fish are investment in shelf-life extension and quality assurance technology, cost efficiency, and organizational experiences able to neutralize the competitive pressure from rivals while taking advantage of opportunities opened by new resources and possibilities offered by growing markets. In this way the matching of value chain conventions is obtained.

Specialized marketing organizations for fresh, high-quality fish products require controlled access to the fresh catch as well as skilled staff trained to manage products quickly through the value chain within a limited time frame. Because fresh fish is highly perishable, it is difficult to maintain control by relying on legal product guarantees. Trust between suppliers and buyers, built up over time, and common product-quality definitions are therefore very important for controlling transactions in the fish marketing chain and to obtain a real matching of conventions (Dulsrud 2002). Establishing such control over the supply chain enables a stable supply of consumer-oriented products, but it is highly constrained by other types of governance conventions of the value chain, like fish quota and harvest management.

13.8 The Influence of Ecological Conventions in Resource Management

Access to and control of fish quotas and market outlets are major barriers to entering or expanding entrepreneurs seeking to build value chains in fisheries. Both fish quotas and market access are regulated and managed by a range of the CIDPEC governance conventions.

The formal and informal management of the common governance conventions takes place in networks of specialists who have their own objectives and interests in the outcome. For example, the clear objective of governmental officials is to interpret and implement legal regulations on resource management and marketing. However, governments depend on help and advice during the implementation process, which opens the way for special interests and professional groups, with their own specific business objectives, to participate. In the Norwegian fisheries, two such regulatory arenas are highly visible: regulation of fishing quotas (ecological conventions) and food safety regulations (civic conventions). Although government agencies manage both these arenas, they do so in close consultation with advisory committees, comprising representatives from industry, scientific institutions, government agencies, and NGOs (Kooiman 2005). Professional guilds specializing in topics such as technology, environment, and consumer issues dominate other, informal, arenas. Furthermore, interest groups in Norway such as fishers' and fish processors' associations manage their own networks through annual meetings and forums for technology platform considerations; among these are the 'fillet forum', 'conventional product forum', and 'pelagic forum', in which the members discuss and coordinate common issues. Research and development projects organized by research institutions and development funds nationally and internationally constitute other examples of arenas where shared conventions are developed and maintained (Whitley 1999).

Firms' participation in these networks facilitates the exchange of information important in developing competitive strategies related to expectations in the relevant environment, and doing so ahead of competitors (Trondsen 1985). Politically strong groups in the fishing sector, such as the Norwegian fishers' association, which represents traditional industrial interests, dominate these governance networks and are a powerful force in reinforcing the traditional, and sometimes outdated, conventions that become barriers to alternative market adaptation conventions (Lindkvist and Sánchez 2008; Trondsen Chap. 9).

13.9 Industrial Harvesting Conventions

Regulation of fish harvesting conventions affects both the ecological and the economic sustainability of fish stocks, the use and production of harvesting technology, and the size and allocation of the economic resource shared among the participant fishers and fish processors. According to Roman law principles—embedded in most European inspired national legal systems today—the fish resource belongs to the commons. Private ownership begins when the fish is caught; governments, on behalf of the commons, have the legal right to manage harvesting of the fish stocks inside a territory of 200 nautical miles, according to the UN Convention on the Law of the Sea.[2] Fish stock harvesting is therefore regulated by political and legal conventions operating under national laws, with such regulation influenced by all governance conventions related to fisheries.

Regulations of the fishing fleet in Norway have, for example, opened up for development of a deep-sea trawler fleet already in the 1950s, which later became a major driver of the restructuring of the fishing industry. These regulations were

[2] Article 61: 'The coastal State shall determine the allowable catch of the living resources in its exclusive economic zone... taking into account the best scientific evidence available ... [to] ensure through proper conservation and management measures that the maintenance of the living resources ... is not endangeredby over-exploitation. ... Such measures shall also be designed to maintain or restore population of harvested species at a level which can produce the maximum sustainable yield, as qualified by relevant environmental and economic factors, including the economic needs of coastal fishing communities.'

justified by the processing industry's need for a more continuous supply of raw materials, independent of fishing seasons when the coastal fleet was in harvesting positions. Trawling technology has always been ecologically controversial because of the trawl's poor effectiveness in fish size selection and the damage it does to the seafloor, in contrast to the passive fishing gears used in coastal fisheries that is harvesting bigger fish. The deep-sea trawler fleet catches more fish far from the coast during bad weather in autumn, when the cod are spreading out in a wide feeding area in the Barents Sea, but it has the same seasonal pattern of landings as the coastal fleet. The trawler fleet is much more capital intensive, which has led to industry lobbying for restructuring and concentration of the limited fishing rights, at the expense of the more environmentally friendly coastal fishing fleet, as shown below. Allocation of fish quotas to trawlers that make longer fishing trips has reduced the quality of the landed fresh fish; this in turn has limited suppliers' access to fresh fish markets with the highest prices and encouraged round freezing on board vessels for later thawing and processing to clipfish and lower-quality refrozen fish fillets (Trondsen, Chap. 9).

13.10 Production and Distribution Conventions as Part of Industrial Conventions

The ecological environment has a strong direct influence on commercial practices in the value chain, as most codfish are available in the first half of the year, whereas market demand is stable year-round or has a seasonality that differs from that of supply. Various regional commercial and industrial conventions, including preservation, storing, and product development downstream in the value chain, have been developed to balance demand pressure and availability of supply. For example, the demand for dried and salted cod has traditionally been highest in the period from December (Christmas) to March/April (Lent and Easter) because of Catholic traditions. The salted cod and dried stock fish industries have had the capacity to purchase huge quantities of big spawning cod from the fleet during winter for salting, drying, and warehousing, to be sold later at the end of the same year and the beginning of the next year, when consumer demand again peaks.

This large cod, mainly caught using selective fishing gear (long line, nets, and jigging), has a natural competitive advantage in the global market over comparable whitefish species such as Alaska pollock (Northern Pacific) and pangasius (Vietnam), which do not have the same fillet thickness so highly valued in the market.

The smaller cod landed mainly by trawlers has traditionally been processed into frozen fillet blocks for further processing into fish sticks or used directly as consumer products such as natural fillets for year-round distribution in frozen food supply chains. However, competitive price pressure intensified with entry into international supply chains of the same-sized frozen fillets of Alaska pollock and pangasius, which has led to the closure of many of Norway's traditional fish fillet factories.

The Norwegian industry responded to the competitive pressure from substitute frozen fillets by diversifying into the production of more fresher specialized products, such as loins, in land factories and by cutting costs by freezing the catch on board trawlers for direct export to other processors, despite the export price for frozen cod being much lower than that for fresh products (Table 13.1). The fresh fish loins are exported and distributed through the major supermarket chains in Europe. Part of the round-frozen cod processed on board the vessels in the Barents Sea is exported directly to low-wage countries such as China and Portugal for further processing into fillets and salted cod products, and part is sent to the clipfish industry in Northwest Norway (Møre region), at the expense of the north Norwegian fish-processing industry.

Iceland's industry had a different strategic response to the increased competitive pressure from the entry of Alaska pollock and pangasius into the international market. Despite similarities between the fishing industries in Norway and Iceland, their different responses to the new competition took them along market specialized development paths, driven by different domestic business orientations. The consequence has been industrial conventions with a different content than the Norwegian. The main difference was that the Icelandic industry responded to the challenge by developing and introducing into the Spanish market a new light salted frozen fillet product (Trondsen, Chap. 9), Lindkvist et al. 2008). In developing this new specialized supply chain and offering light salted frozen products, the Icelandic industry was taking advantage of a new market opportunity, applying new additive technologies and relying on fresh fish landings—in stark contrast to the Norwegian approach of freezing unprocessed fish on board trawlers and sending it for processing in areas with cheaper labor while closing their own frozen fillet processing factories. These differences in path development are related to differences in regulation and governance by the authorities as part of civic conventions.

13.11 The Influence of Firsthand Exchange Conventions

The Icelandic case demonstrates that regulatory changes leading to the introduction of firsthand sales auctions as a market influenced domestic industrial convention were important in placing structural pressure on the industrial raw fish input, thus forcing commercial changes and conventions further down the value chain. By contrast, the Norwegian domestic approach to firsthand sales created protection against market-oriented price variations and structural changes further down the value chain.

Under the auction exchange mechanism, firsthand buyers must offer fishers competitive prices for all the landed fish qualities they want; however, this can only be done by maintaining specialized market-oriented production conventions for all of the high-priced products that satisfy new market trends for convenience and flavour.

In northern Norway, close to the main cod fishing grounds, a regulated price system is in place, whereby fishers are guaranteed administrative set minimum prices for all fish they land. Firsthand buyers do not have to purchase any fish, but they do have to purchase all fish the vessel offer in a landing at a minimum price even if they are interested in only part of the catch, independent of the seasonal supply–demand balance. The main reason for the Norwegian system of a regulated minimum price, which was introduced in the 1930s, was to protect fishers against the variation in prices due to the distance in space and time between fish landings and consumption areas. This system is protected by the Raw Fish Act and managed by the fishers' sales organization (Hallenstvedt 1982).

The fish auction as part of industrial and domestic conventions in the EU and Iceland differ from those in Norway. Both have a legal minimum price system, involving a shadow price that must be met if a sale at auction is to be completed. However, in the EU and Iceland, this shadow price is decided by the government, not by the fisher-controlled organizations as in Norway. The result is that minimum prices in Norway are significantly higher than those in the EU and Iceland. Fishers in the EU and Iceland are under greater pressure to follow market conventions if they are to obtain the highest prices for all species in their landings—an opportunity presented by the fish auctions. About 17–20 % of all cod landings in Iceland are sold through the fish auction network, which relates very closely to the figures for fresh fish exports (Table 13.1). This network is very efficient at communicating the prices in the fresh fish market for each fish quality and species, which improves the fishers' knowledge base and thus supports their market-oriented behavior. Furthermore, according to a union agreement in Iceland, all other vessels, such as deep-sea longliners and trawlers with vertically integrated fish-processing plants, must pay their fishers a share of the catch value, with the value of each species and fish size calculated as 76 % of the fish auction price (Trondsen and Arnarson 2010). In this way, Iceland's regulations and management of the fish auctions place market-oriented competitive pressure on the entire value chain, thus motivating the development of commercial and industrial conventions driven by the fresh fish market with the highest prices. This supply chain structure supports market-oriented commercial conventions for firms operating in the final consumer market; as it enables them to always purchase fish at auction that satisfy the end consumers' quality conventions. This opportunity is not available to Norwegian exporters, as they are more dependent on exporting landed fish, which is quality differentiated according not to market preferences but rather to catch pressure and landing regulations.

Norges Råfisklag, the Norwegian fishers' sales organization that represents the governance of firsthand sales conventions in northern Norway, tried some years ago to implement a fish auction, but without success. The fish processors did not like the auction's effect on the fish market and colluded against it. However, Norges Råfisklag did succeed in introducing an auction for whole fish frozen on board vessels. Over time, this auction has become a price leader in the raw fish market, giving price incentives for more freezing of catches on board vessels, at the expense of the land processing and fresh export industries, which are dependent on fresh landings (Helstad et al. 2005).

The undermining of attempts to introduce fresh fish auctions and thus new industrial conventions and regulations in Norway demonstrates a lack of strong market pressure to increase the value of landed fish. The industry satisfied its economic needs by controlling limited resources even without specializing in consumer market orientation.

13.12 The Economic Influence of Governance of Fishing Effort

The sustainable supply of wild fish for the global commodity market is limited. The balance between total global supply and the demand for scarce fish species is an important structural economic constraint on value adding in the fishing industries that influences the level of profit satisfaction. Limits to growth on the supply side are a characteristic supply barrier in natural resource-based value chains.

At the other end of the value chain with its market and consumer conventions, the demand for fish products is increasing, driven by growing consumer awareness of the health benefits of consuming seafood combined with general population growth and increasing total consumer income (see, e.g., Trondsen et al. 2004a, b; Swartz et al. 2010). With the combination of an increasing demand and a stable or even decreasing supply from wild fish stocks, fish is increasingly becoming a scarce good with prices for the raw materials growing, independently of the value chain's degree of market orientation.

Variation in the pattern of fish landings and value-adding opportunities are also affected by regulation of the fleet's catching and firsthand sales organization. The introduction of individual fishing quotas for different species concentrated among a limited number of vessels, in a bid to limit the catch to a sustainable harvesting level, has led to the concentration of catching of each fish stock in periods of the year when the margin between sales value and catching cost is highest. Regulations that grant each vessel an individual quota for each fish species motivate fishers to harvest only in the peak season, using the most efficient equipment for each species, even if supply pressure and quality reduce the landing price compared with seasons with lower catch rates and higher catching costs. In Norway, this landing behaviour favours the commercial conventions that dominate the drying and salting industry, which can absorb huge amounts of fish in a short period of time. Because all salted fish goes into warehousing to be sold to consumers several months after landing, there is no direct consumer reaction to the prices of the quantity supplied during these short winter harvesting seasons. Consumer reactions come some months later, which may cause price fluctuations that affect the profits of traders who accept a significant warehousing risk each year.

Therefore, all participants in the industry have a strong interest in allocation of restricted fish quotas, which has the effect of creating competition between industry participants to achieve the highest possible share of the valuable fish quota and landings. These production-oriented industrial conventions, which focus on competition for quota and landings in the raw fish market, contrast with more market-oriented conventions, which profit from improving market satisfaction by offering better value-added products and services (Trondsen and Johnston 1998; Grunert et al. 2010). The Norwegian fishing industry decision to specialize in production-oriented or industrial conventions is an indication that managers perceive that marginal improvements in profit satisfaction from controlling fish quotas and landings are greater than those derived from specializing in consumer conventions. Profit satisfaction rather than profit maximizing is according to Simon (1957) and Cyert and March

(1963) the dominating aim for business organizations, which means that interest in changing behaviour is removed when profit satisfaction is achieved, even if there is untapped value adding possibilities in the market place.

A comparison of the Norwegian and Icelandic fisheries management strategies gives some insights into the factors influencing the balance between specializing in supply-oriented and market-oriented industrial conventions in a value chain.

Both countries have cod fisheries regulated by individual quotas. However, whereas Icelandic quotas can be more flexibly traded between firms as individual transferrable quotas, which can be subject to permanent or temporary transfers via leasing contracts, quota transfers in Norway are mainly linked to the transfer of vessels between firms. In addition, Iceland has a fresh fish auction linking the fresh fish value chains to the consumer market, which places market-oriented competitive pressure on all links in the value chain. By contrast, in the Norwegian cod fisheries, which are dominated by minimum price regulation, management convention limits rather than increases market pressure on fishers and the firsthand supply chain. The lack of flexible quota transferability in the Norwegian system is also a barrier to flexible market adaptation. Furthermore, fish quotas in Norway are greatly controlled by fishers that have no investment in land production facilities, whereas in Iceland, quotas are mainly owned by vertically integrated companies with interests in both processing plants and fishing vessels. The increasing shortage of fish quotas relative to demand has driven up the value of the quotas in most countries. Individual quotas allocated by the government for free, as a windfall for lucky fishers, have become the most important economic asset in the Norwegian industry, generating a much higher profitability than investment in more specialized value-adding market and consumer conventions. The legal divide between the Norwegian fishing vessel companies that own fish quotas and the land-based processing sector has motivated specialization in industrial conventions where the fleet invests in onboard freezing of fish for direct export for further processing in low-wage countries, rather than investing in more value-adding market-oriented production involving land-based processing plants specializing in fresh fish processing.

13.13 Summary and Conclusions

This paper has identified common constraining characteristics for market adaptations of the value chains for wild harvested seafood. Most important is the seasonal mismatch in supply and demand and distance in space between the harvest and the consumption regions and the choice of preservation methods. The study also shows how value chains conventions, geographically rooted in harvesting and processing regions upstream, are constraining and facilitating differences in market adaptation downstream. Most important is the regulation of fish quota allocation and method of first hand sales of the catches.

The paper shows that the cod value chain in Norway has, compared with that in Iceland, demonstrated less capacity to take advantage of the increasing demand for highly valued, fresh, good-quality fish products. Most of the commercial and value-adding conventions applied in the Norwegian fish industry are production oriented dealing with managing lower-valued, preserved, semi-processed products for further processing further down the value chain. Iceland has on its part managed to develop value chain conventions which are gaining a higher market oriented value adding compared to its Norwegian competitors. The divergence between consumer markets' preferred supply conventions and the actual export conventions favouring lower-value, preserved products is partly explained by fluctuations in the ecological environment and partly the conventions shaping regulations and behavior of fish harvesting as the first stage in the value chain. The government's allocation of individual fishing quotas and guarantees under the Raw Fish Act of minimum prices for all landings independent of market adaptation have oriented the controlling industrial conventions towards harvesting rather to market adaptation.

The findings also underscore the fact that governmental regulation of the fish harvesting sector upstream has an impact on the market adaptation downstream, even if it's limited to managing sustainable fish stocks and quota allocation among stakeholders. Improving the market oriented value adding capabilities on behalf of the public trust of such a natural resource based industry requires value chain governance conventions which optimize the added value from the demand and supply driven transactions.

Acknowledgements The author is thankful for critical comments and recommendations of an earlier draft of the paper from my colleagues Knut Lindkvist and José Luis Sànchez Hernàndez.

References

Andersen MA (2009) Regionalizing global trade patterns, 1981–2001: application of a new method. Can Geogr 53(1):20–44

Bestor TC (2004) Tsukiji. The fish market at the center of the world. University of California Press, Los Angeles

Cano CR, Carrillat FA, Jaramillo F (2004) A meta-analysis of the relationship between market orientation and business performance: evidence from five continents. Int J Res Mark 21:179–200

Crevoisier O (2004) The innovative milieus approach: toward a territorialized understanding of the economy? Econ Geogr 80 (4):367

Cyert RM, March JG (1963) A behavioural theory of the firm. Prentice Hall Inc, New Jersey

Day GS (1994) The capabilities of market driven organizations. J Mark 58(4):37–52

Dulsrud A (2002) Tillit og Transaksjoner. En kvalitativ analyse av kontraktsrelasjoner i norsk hvitfiskeksport. National Institute for Consumer Research, Report 2, Oslo

Fisher ML (1997) What is the right supply chain for your product? Harv Bus Rev 75(2):1005–1116

Gereffi G, Humphrey J, Sturgeon T (2005) The governance of global value chains. Rev Int Political Econ 12(1):78—104

Grunert KG, Trondsen T, Campos EG, Young JA (2010) Market orientation in the mental models of decision makers. Two cross-border value chains. Int Mark Rev 27(1):7–27

Hallenstvedt A (1982) Med lov og organisasjon: organisering av interesser og markeder i norsk fiskerinæring. Universitetsforlaget, Oslo

Helstad K, Vassdal T, Trondsen T (2005) Price links between auction and direct sales of fresh and frozen fish in North Norway (1997–2003). Mar Resour Econ 20:305–322
Kooiman J et al (eds) (2005) Fish for life: interactive governance for fisheries. Amsterdam University Press, Amsterdam
Kurlandsky M (1999) The Basque history of the world. Alfred A. Knopf, Toronto
Lindkvist KB, Sánchez J (2008) Conventions and innovation: a comparison of two localized natural resource-based industries. Reg Stud 42(3):343–354
Lindkvist KB, Gallart-Jornet L, Stabell MC (2008) The restructuring of the Spanish salted fish market. Can Geogr 52(1):105–120
Myrland Ø, Johnston RS, Lund E, Trondsen T (2000) Determinants of seafood consumption in Norway: lifestyle, revealed preferences, and barriers to consumption. Food Qual Prefer 11(3):169–188
Narver JC, Slater SF, MacLachlan DL (2004) Responsive and proactive market orientation and new-product success. J Prod Innov Manage 21:334–347
Porter M (1990) The competitive advantage of nations. The Macmillan Press Ltd, London
Sánchez-Hernández JL (2011) The food value chain as a locus for (Dis) agreement: conventions and qualities in the Spanish wine and Norwegian Salted-Cod Industries. Geografiska Annaler: Series B. Hum Geogr 93(2):105—119
Scholderer J, Trondsen T (2008) The dynamics of consumer behavior: on habit, discontent, and other fish to fry. Appetite 51:576–591
Simon H (1957). Administrative behavior. Free Press, New York
Stevens JS (1980–1981) The public trust: a sovereign's ancient prerogative becomes the people's environment right. The University of California Davis Law Review 195–210. http://heinonline.org/HOL/LandingPage?handle=hein.journals/davlr14&div=16&id=&page=
Swartz W, Rashid Sumaila U, Watson R, Pauly D (2010) Sourcing seafood for the three major markets: the EU, Japan and the USA. Mar Policy 34:1255–1373
Trondsen T (1985) Industriell Innovasjon. En studie av institusjonelle forutsetninger. Dr. Scient. thesis, Institute for fishery science, University of Tromsø
Trondsen T (2012) Value chains, business conventions, and market adaptation: a comparative analysis of Norwegian and Icelandic fish exports. The Canadian Geographer/Le Géographe canadien 56(4):459—473
Trondsen T, Arnarson I (2010) The salted fish project. report from iceland survey. Working paper, The Norwegian College of fishery Science, University of Tromsø
Trondsen T, Johnston RS (1998) Market orientation and raw material control. J Mark Focus Manage 3:193–210
Trondsen T, Young JA (2006) The role of fish auctions in value adding in fish marketing chains. In: Ashe F (ed) Primary industries facing global markets. The supply chain and markets for Norwegian food. Universitetsforlaget, Oslo
Trondsen T, Helstad K, Young J (2003a) Market-oriented regional fisheries management? An analysis of four fish regions in the North Atlantic. Ocean Coast Manage 46 (9–10):917–941
Trondsen T, Scholderer J, Lund E, Eggen AE (2003b) Perceived barriers to consumption of fish among Norwegian women. Appetite 41(3):301–314
Trondsen T, Braaten T, Lund E, Eggen AE (2004a) Health and seafood consumption patterns among women aged 45–69 years. A Norwegian seafood consumption study. Food Qual Prefer 15(2):117–128
Trondsen T, Braaten T, Lund E, Eggen AE (2004b) Consumption of seafood—the influence of overweight and health beliefs. Food Qual Prefer 15(4):361–374
Whitely R (1999) Divergent capitalisms: the social structuring and change of business systems. Oxford University Press, Oxford

Chapter 14
An Analysis for a Norwegian Recapturing of a Salted Fish Market in Spain

Heidi Bjønnes Larsen, Knut Bjørn Lindkvist and Torbjørn Trondsen

Abstract This chapter sums up the drivers and barriers for Norwegian producers in the Spanish market for salted fish. Four decisive changes in the lasting deep structures in the connecting value chain are identified in this book: (i) In the declining demand for traditional salted fish. (ii) In the composition of demanded products in the market. (iii) In the geographical patterns of the dominant consumer conventions and finally, (iv) In the lack of market oriented adaption of Norwegian-Spanish value chain conventions. The chapter applies critical realism as a platform to explore strategies to overcome the trade barrier in order to adjust the current causal deep structures in the value chain. Based on a thorough empirical study the chapter discusses possible strategies for producers to create interactive convention platforms between production knowledge and market knowledge in order to develop competitive, sustainable and profitable marketing strategies.

Keywords Critical realism · Deep market structures · Structural changes of products · Geography and conventions · Market strategy · Market adaptation · Market integration · Quality standards

14.1 Introduction

Salted codfish is a very important Norwegian export product that accounted for 48 % of the total cod fish landings in 2012. About 40 % of the Norwegian export of salted fish (about 47,000 t to a value of 1.75 billion NOK) was exported to Portugal and Spain in 2012 (Source: Seafood Norway, special delivery of fishery statics to The Salted Fish Project). Salted fish is an important product on the Iberian Peninsula. In

H. B. Larsen (✉) · K. B. Lindkvist · T. Trondsen
Department of Geography, University of Bergen, Bergen, Norway
e-mail: heidi.bjoennes@gmail.com

K. B. Lindkvist
Finnmark University College, Finnmark, Norway
e-mail: Knut.Lindkvist@geog.uib.no

T. Trondsen
Norwegian College of Fishery Science, University of Tromsø, Tromsø, Norway

K. B. Lindkvist, T. Trondsen (eds.), *Nordic-Iberian Cod Value Chains*,
MARE Publication Series 8, DOI 10.1007/978-3-319-16405-2_14

both Spain and Portugal, salted cod (or Bacalao/Bacalhau) is considered as a distinct food category, different from other kinds of fish. Salted cod is sold in special shops together with salted herring, pilchard or anchovy and other pickled products and has dedicated stalls in the marketplace and in supermarkets. Bacalao has a unique set of recipes and plays a central role on religious occasions such as Christmas and Lent. Despite the significance of Bacalao, consumption of the traditional salted fish product on the Iberian Peninsula has decreased with the development of new products.

The salted fish trade between Norway and Spain has been important since the sixteenth century (see Chap. 3). Norway has had an abundance of fish to offer and has been a central actor in the Spanish market, but since 1998, the Norwegian export of 13,000 t of salted fish products to Spain has declined. The Norwegian market share of salted fish products in Spain has declined as well. More recently, the annual export of salted fish to Spain has been stable at 5000–7000 t, which is half of what it was in previous years. The Norwegian export to Portugal was 40,000 t in 2012. However, in this chapter, the main focus will be on Spain, as there is a possibility that the Norwegian salted fish industry can recapture this market.

The decline in Norwegian exports to Spain since 1998 is partly attributable to the restructuring of the market and the demand for new convenience products containing less salt. The recent decline in Norwegian exports is also partly attributable to the spread of phosphate-injected products during this same period. The colour of the product further differentiates producers and countries that export to Spain, as do production practices such as treatment of the fish prior to the salting process, how long the fish has been cured and whether chemicals have been added during production. The Spanish producers have preferred to work with Icelandic exporters, who deliver a whiter salted fish than the Norwegian exporters. The Icelandic–Spanish relationship has led to a basic change in Spanish consumers' product preferences and a new geographical pattern of quality conventions. Catalonia is a trend-setting region with a strong economic position in Spain. For the past 100 years, Catalonia has had a clear preference for absolutely white and firm-bodied salted fish (see Chap. 3). Icelandic producers took advantage of this demand when they entered the market and began working with producers in Catalonia. The conventions and the quality standards for salted fish products (white, "clean fish") have spread from Catalonia and to the rest of Spain (see Chap. 6).

Why has Norway lost its influence in the Spanish market? Several scholars have offered explanations (Lindkvist et al. 2008; Lindkvist and Sánchez 2008; Lindkvist 2010; Sánchez-Hernández 2011; Trondsen 2012). Norwegian producers do not produce for a specific market but for several markets that are able to absorb variations in supply from the fish landing. The producers mainly focus on short-term price variations caused by fluctuations in supply; they are price oriented rather than being attuned to market relations. However, business relationships are important in Spain, and the Norwegian producers' focus on short-term prices may hinder their deeper integration into the market. Norwegian production conventions are not suited to Spanish market conventions (such as the Norwegian regulation that prohibits the use of phosphate injections). Thus, Norwegian producers are not willing or able to produce what the Spanish customers want. Norwegian salted fish is a generic product that is offered to several markets. The export strategies are mainly oriented towards the customers' willingness to pay the highest prices in the spot market. Other parts of the Norwegian institutional framework do not encourage the consideration of individual and special market needs. In contrast to Iceland, Norway has not established market differentiated quality standards for salted fish products. The system of first-hand sales of fish, protected by the Norwegian Raw fish Act[1], does not facilitate market-oriented product specialization. Even if fish processors only want a single species or grade of fish to process for their customers, according to the Raw Fish Act, they are obligated to buy the entire mix of fish species and qualities available in the fresh catches offered by the commercial fishing companies. Landings of fish that are frozen at sea are excluded from this regulation, and thus buyers may purchase single grades or species of fish from frozen storages. Lower-grade raw fish that enters the production chain may be used in low-quality products downstream in the value chains to the markets. Market-oriented differentiation in fish processing is constrained, therefore, by the obligation to purchase raw fish.

To secure a "clean natural product", Norwegian food regulations prohibit the use of phosphate as an additive in salted fish, which prevents Norwegian salted fish producers from participating in the market for white, phosphate-injected fish. The purpose of this discussion is to contribute to a scientific analysis of market development by isolating the mechanisms and the deeper causal structures that may have led to this situation for Norwegian fish producers. According to Sayer (2006), the question to ask with respect to causation in this matter is: What has caused the patterns we see in the salted fish markets and value chains? To isolate the often numerous mechanisms that can be responsible for concrete events, Yeung (1997) suggested that formal (statistical data), social (interview data) and analytical (discussion) simplification criteria (at appropriate levels of abstraction) be used.

In the following sections, we will first give a theoretical overview of our analytical perspective. In the Sect. 14.2, we will go deeper into the long-term development of the market for salted fish in the Iberian Peninsula as a whole, and in Sect. 14.3, we outline the drivers for the value chain's market adaption. In the Sect. 14.4, we discuss strategies for filling the gap between the trend in market preferences and the Norwegian value chain development.

14.2 Analytical Framework

Our time perspective is 20 years. The changes we are concerned with are not only those in the salted fish market generally but also basic changes in society, value chains and product structure, which are connected to product demand. These trends

[1] http://www.lovdata.no/cgi-wift/wiftldles?doc=/app/gratis/www/docroot/all/nl-19511214-003.html&emne=R%C5FISKLOV*&.

accumulate and create geographical patterns in material production and sales structures, and thereby change social practices in given regions. The deep institutional and economic structures that constrain participants' room to manoeuvre in the Norwegian and Spanish value chain are in the same category as long-term trends. "Deep structures" refer to the study of the underlying reality of the value chain, as well as to basic categories and relationships (Groff 2000). Structures are "sets of internally related objects or practices"(Sayer 1992, p. 92) as institutions tied together and defined according to each other. These structures refer to how the trade is practised and organized by the participants: Norwegian fisheries, producers, distributors, Spanish producers, retailers and last, but not least, consumers.

The background for our approach is found in critical realism as a philosophy of science and an analytical platform (Sayer 1992; Yeung 1997; Groff 2000). The research conducted in the Salted Fish Project has shown that socio-economic structures in the form of accepted practices and production conventions existed prior to the current era of production in different plants. The socio-economic institutional structures are related conventions or social decisions and practices that have a critical influence on production, market adaptation and innovation. These social structures have been supported by material structures in the form of a legal framework and production facilities with technological devices. That something is a deep structure in this instance means that over a long period of time, with more or less permanent institutions (such as conventions), it has had a decisive and lasting influence on the choice of actions or adaptations made by companies that produce or sell salted fish. In our further discussion, we will consider the course of innovative changes in this perspective.

The course of changes outside the production facilities can be such that participants in the process are required to submit to the changes or to be forced out—an either/or situation. The changes are economically necessary for those who wish to participate. A wise market adaptation for participants in a relevant production system will be to adjust to the changes so that the basis for production is maintained and improved, often through development of new products. In this situation, incorporated and commercialized changes function as mechanisms that help firms to stay in contact with the market's deep structures.

In our further discussion, we will analyse some characteristics of four "deep" structures, which are linked to mechanisms and events (Sayer 1992). A mechanism, like a regulation, supports the structure, but it may also produce structural change (Sayer 2006). Events that result from structures and mechanisms can be directly experienced and observed, and they may also cause changes in deep structures. These deep structures are past-dependent, but as they also represent changes over years, they affect the Iberian markets' lasting demand structure. The first two deep structures to be discussed are the fundamental demand for salted fish in Spain and the long-lasting composition of product mixes in the market. These two structures are interrelated but will be analysed separately. Demand and the products produced change over time and space, and such changes can be more easily discussed as demonstrations of change in independent deep structures. The third deep structure, the different internal geographical patterns of the fish markets, also appears as an independent

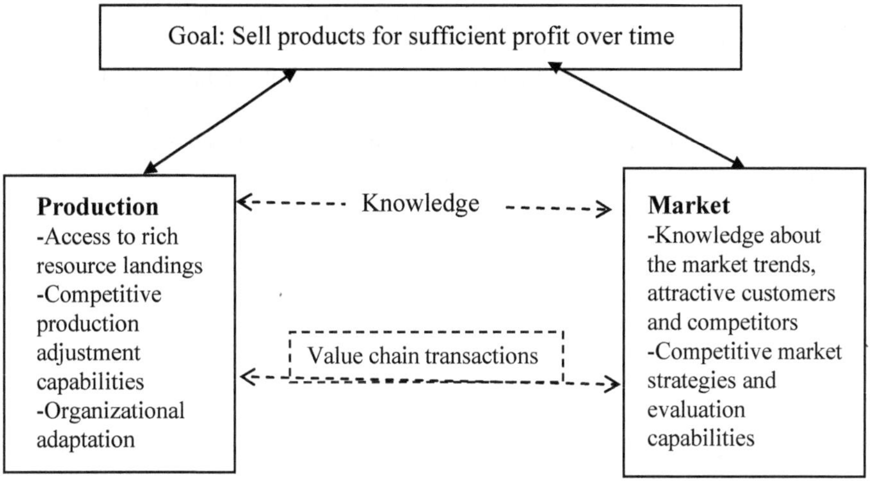

Fig. 14.1 Competitive strategy constraints in the Norwegian industry for salted whitefish. (after Larsen and Lindkvist 2014)

deep structure. Every region can be assumed to have special characteristics that separate it from other regions, but here too, the deep structures relating to geography have changed during the past 20 years. The last deep structure we will discuss is the concrete Spanish–Norwegian value chain for salted fish production, which has been characterized according to the conventions of production and consumption. This value chain has had some "synchronic" or stable characteristics over several generations. As important as the deep structures in shaping trends and structural changes are the mechanisms that "produce" and sustain deep structures. Regulations are tools for changing transactional practices in the value chain. Regulations of the fish quota, first-hand sales transactions and the use of phosphate may all contribute to institutionalization of trade and production behaviours in Norwegian and Spanish firms. Based on the deep structures, we will discuss the Norwegian salted codfish industry's production and export strategies for the Iberian Peninsula. Of course, the fundamental and primary deep structure is the regulated system of capitalism, which spurs the firms to seek profits, both short-term and long-term.

The key to a successful strategy is the ability to find structural causality under the surface of the value chain's deep structure (Porter 1980). Figure 14.1 illustrates aspects of the value chain and how the goal for the Norwegian industry is based on production and market-specific entities and knowledge. The industry needs to be competitive both in the production stages, to gain access to raw materials, and in the market, to attract customers. The production stations need to acquire sufficient fish resources and capabilities for product adjustments and organizational adaptation.

Successful production stations take advantage of knowledge evaluations and signals about target markets, attractive customers and competitors' activities to develop their strategies for competing in the market. Realizing business goals depends upon production, as well as the market and transactions in the value chain. The discussion of possible strategies relies on co-ordination between production and the acquisition of sufficient market values for transaction partners. Our theoretical foundation is in institutional and evolutionary economy, and more specifically in economic geography, theories of internationalization, value chain governance, and the importance of related variety that allows people to learn from each other and, thus, the different knowledge bases of production. The strategies discussed are based on data from both the Norwegian–Spanish salted fish project, reported in the previous chapters, and additional interviews conducted with 75 % of the members of AiB (Associados industrias do bacalhau) in Portugal. Our focus here is on the deep structures and their changes.

14.3 Deep Structural Change 1: Decline in the Demand for Traditional Salted Fish

The production of salted fish has had a long development process. Salted fish has had an important place in Spanish culture since pre-Roman times (Lindkvist et al. 2008). Tuna fish was salted and dried during the hot Mediterranean summers. During Roman times, Hispania was praised for its high-quality salted fish, especially tuna, and it was considered one of the most important areas for producing and exporting salted fish. After a decline in salted fish production during Muslim domination, the tradition re-emerged in the thirteenth century but with a new focus on cod (Gallart-Jornet et al. 2005). Prior to the twentieth century, salted codfish was imported to Spain from Canada, Great Britain, France and Norway. Later in the twentieth century, salted fish was mainly supplied by the Nordic countries, especially Norway, and Spain's own trawler fleet (Lindkvist et al. 2008). Today, there is no need to salt fish for preservation, but salted fish still prevails as a distinct product, and it is deeply appreciated by consumers on the Iberian Peninsula. Portuguese producers emphasized the importance of bacalhau in Portuguese culture and food traditions. The fish is dear to the Portuguese people and is a long-standing staple of the Portuguese market. Thus, the product appears to be linked to a deep synchronic food culture. Even with newly introduced production processes, salted fish is still the product platform for many other products that have evolved in related industries. The reason is that salted fish has played an important part in Spanish and Portuguese history. Richter-Hanssen (see Chap. 3) explains the central role of salted fish during the harvest of wine and grain. Historically, in Navarra, Aragón, Castilla, Rioja and Galicia, large work-forces needed salty, nutritious and convenient food to sustain them and replace the salt they lost while working in the warm fields. Norwegian fish was sold to consumers living in the country-side and the labouring class in the cities. Norwegian fish was reasonably priced and hence popular among consumers with little money. Wealthier citizens preferred the tastier, whiter and better-looking fish from Iceland or the Faeroe Islands.

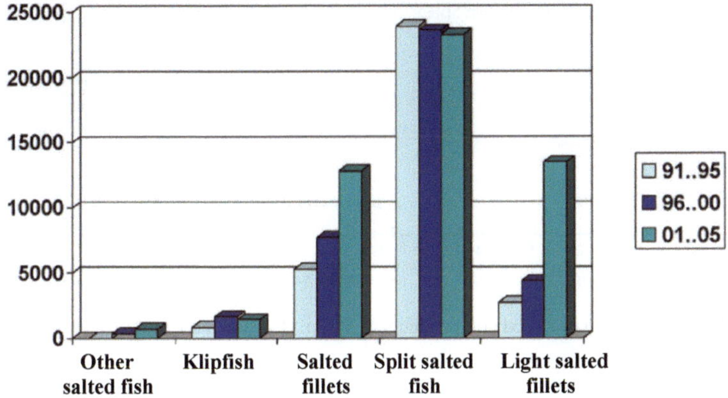

Fig. 14.2 Structural changes in the Spanish importation of salted fish, 1991–2005. (after Lindkvist et al. 2008)

Currently, Spain is the largest importer of seafood in Europe, and the fourth largest importer in the world (Sleipnes and Johansen 2012). Despite this strong position, the Iberian Peninsula has been importing less of the traditional salted fish products in the past 20 years. Spain's total importation of such products increased from 21,000 t in 1998 to 41,000 t in 2006, and then decreased to 31,000 t in 2008. In the same period, Norwegian exports to Spain decreased from 13,000 t in 1998 to about 6000 t per year in recent years. The Portuguese annual importation of salted cod and klipfish decreased from 80,000 t at the beginning of the 1990s to 55,000 t in 2012. The Norwegian export of salted fish to Portugal in 2012 was 40,000 t, which required 38 % of the total Norwegian cod landings.

14.4 Deep Structure 2: The Change in Composition of Products in a Market

Another deep past-dependent structure is the composition of products made from salted cod but still with changes over the years.

Figure 14.2 shows the changes in salted fish imports to Spain from 1991 to 2005. The importation of traditional products decreased, whereas salted fillets, and especially light salted frozen fillets, increased. The main product for Norwegian producers is wet salted and dried salted cod (klipfish), both of which are raw material used for further processing in numerous other products in Spain. The decline is a function of a deep, past-dependent structural trend that has had a decisive and lasting influence on the companies' production profiles.

The traditional technology was salting and drying, which protects against microorganism growth. In the past, the main product was the hard, salted, whole split fish, and klipfish. The fish could be stored for months without being destroyed and

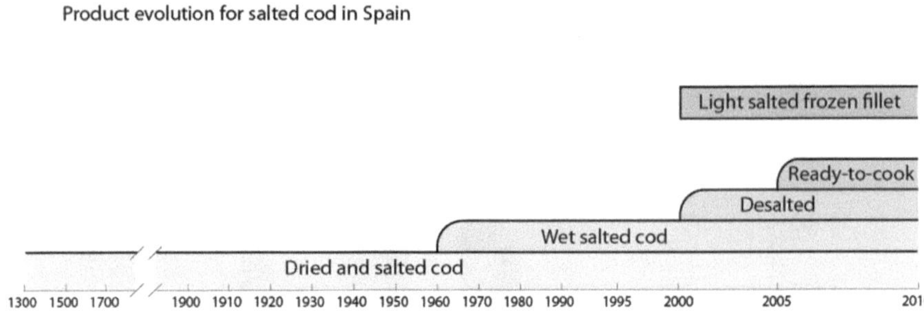

Fig. 14.3 The evolution of salted cod products in Spain. (Larsen and Lindkvist 2014)

was a staple food for numerous communities, not only in the Iberian Peninsula. When cooling technology became widespread in the 1960s and 1970s, the need to salt, and to dry the salted fish, for conservation purposes declined. Through the use of cooling technologies, many Norwegian producers were able to sell a lighter wet salted cod directly to distributors and to by-pass the klipfish processors. In the 1990s, Icelandic producers adapted brine injection machines from the meat industry to the fish salting process, using brine that contained phosphate (Lindkvist et al. 2008). The result was light salted frozen fillets that contained only 2 % salt and tasted more like fresh fish. The product is not cured, and therefore it is in a different category, but it was developed based on the traditional demand for cod with a salty taste. Xie (see Chap. 10) shows that frozen light salted fish and consumer packages are growing in demand and that to a certain degree, light salted frozen fillets from Iceland have become a substitute for the traditional salted cod products from Norway. Iceland created a separate statistical category for light salted frozen fillets in 2007; previously, they had been categorized as frozen fillets.

The traditional product, which is still preferred by many, needs to be desalted and rehydrated before it can be eaten. This was traditionally done in homes and restaurants, but with the growing consumer trend for convenience, a demand for desalted products emerged. The Spanish producers took advantage of this demand and offered frozen or chilled ready-to-cook and ready-to-eat desalted portion products.

Figure 14.3 shows the evolution of salted cod products available to the Spanish market. All the current salted cod products have evolved from a traditional demand for klipfish and salted whole split cod.

The introduction of phosphate has gradually changed Spanish consumers' perceptions of quality. Larsen (see Chap. 6) and Espinosa and Martínez (see Chap. 5) showed how the majority of Spanish consumers have come to prefer a whiter codfish, less cured and with a softer texture. The supply systems in Spain have changed to meet these evolving consumer preferences. Larsen (see Chap. 6) concluded that current quality standards probably will be the basis for future production systems because of the current standards' domination of industrial conventions.

14.5 Deep Structure 3: Geographical Patterns of the Dominant Conventions

The salted fish value chain, including consumption, has a spatial distribution. Socio-economic mechanisms transform regions into matrices of economic development (Crevoisier 2004, p. 367). The regional basis of food consumption and distribution are, therefore, deeply rooted historical structures that are past-dependent and caused by lasting economic and cultural imprints that change slowly over time.

The 1990 map of salted fish colour preferences illustrates a heterogeneous Spanish market, with a slight majority towards the golden colour. The 2010 map shows that the market has become more homogeneous, with a clear majority demand for a white product (see Chap. 6). Although white colour has become a general trend in the salted cod market, there is still variation in this market, with preferences for different types of qualities corresponding to different geographical structures. Espinosa and Martínez (see Chap. 5,) clearly demonstrate that quality features and conventions are geographically embedded in Spanish submarkets. A certain diversity in regional preferences provides the opportunity to investigate alternative ways of governing the value chain.

14.6 Deep Structure 4: The Norwegian–Spanish Salted Fish Value Chain

A value chain connecting production and consumption is not linear but circuitous, with feedback loops of money flows (payments for goods and services) and information (demands from customers) (Trondsen 2012; Dicken 2007). The fragmentation of a value chain may hinder the feedback loop to the producers of market information other than prices. The Norwegian–Spanish salted fish value chain is a lasting and synchronic structure that is supported by mechanisms such as laws and regulations, investments, and supply and demand. This way of organizing and operating processing and trade has existed for the last several decades in its general form. The value chain as a structure has had a decisive and lasting influence on the choice of actions or adaptations available to companies that sell commodities in the market. The value chain is organized in what we have called "stations". All the stations are connected through interfacing gates. The main stations in the value chain between Norway and Spain are all filled with production structures of resources and people that follow dominant industrial conventions; they are pulled by demand and pushed by pressure from fish landings. The first cod production station is fish harvesting, supplying fish for processing as the second station. The operation of the harvesting station is constrained on the supply side by regulations in the form of fish quotas, and on the sales side by the Norwegian first-hand sales regulation, which is carried out by fishermen's sales organizations (such as Norges Råfisklag in the north). The Norwegian salted fish producers have several trading gate options, including their Spanish or Portuguese importers and the klipfish processors in Norway. The fish quality in this trade is regulated by the Norwegian food authority. In Spain and Portugal, the fish is further processed (for instance, to klipfish or as wet fish), portion packed and sold to consumers through retailers, supermarkets or restaurants. All of the physical structures associated with these stations function as mechanisms and as barriers or drivers for change.

The salted fish value chain between Norway and Spain consists of many very small firms that are not vertically integrated, as transnational firms are. Increasing vertical co-ordination also increases transaction costs (Gereffi et al. 2005). Co-ordination costs are reduced by repeated transactions, relations, or matching common conventions. As a possible strategy, integration may include these steps. We will also look at other strategies.

14.7 Strategies

14.7.1 Long-Term Goals

A strategy is a determination of the basic long-term goals of an enterprise, and the adoption of courses of action and the allocation of resources necessary to reach these goals (Chandler 1962, p 39). A strategic objective is to maintain or improve performance by exploiting opportunities and neutralizing threats to the organization (Porter 1980). A strategy describes the direction an organization takes to match its resources with its environment to satisfy its stakeholders (Whipp 2001).

Three possible market-oriented, profitable strategies that Norwegian salt fish producers can take for the Iberian Peninsula are discussed in the following sections.

14.7.2 The Current Strategic Practice for Market Adaptation

The current Norwegian strategy is the production and sale of generic products, which provides the flexibility to operate in a number of markets. An evolutionary approach argues that "the explanation for why something exists intimately rests on how it became what it is" (Dosi 1997, p. 1531). Historically, salted fish producers (such as those in the north of Norway) have been part of a value chain dominated by a captive sector and fishing-industry-controlled organizations. The salted fish industry or klipfish industry has been controlled by a number of monopolies. Up to 1789, a Copenhagen monopoly controlled the salted fish trade from northern Norway, and the products were resold as Danish (3). In 1914, during World War I, the Norwegian state formed a monopoly on all sales of klipfish, centralized in the south-western city of Ålesund. From 1971 to 1992, Fiskeprodusentenes Fellessalg exercised a monopoly on the sale of all klipfish, dried fish and salted fish produced between north-west and northern Norway. This previous experience of the fishing industry's control over the value chains through the Fishermen's Law, with protected sales organizations and trade monopolies, may explain the present market-co-ordinated chain and the mistrust among actors in the salted fish value chain (Lindkvist 2010). Taking an historical approach, Richter-Hanssen (see Chap. 3) shows the mechanisms of monopolies and demonstrates how Norwegian production conventions and regulations (differential duties, preferences for Spanish boats and, later on, for monopolies) have affected the evolution of the path-dependent Norwegian value chain. There was a time when the Spanish market was dependent on Norwegian fish, and Norwegian producers had little incentive to satisfy Spanish consumer preferences. The Icelandic value chain, on the other hand, was driven by market conventions (Trondsen 2012). Iceland developed a co-ordinated and integrated value chain with Spain (Richter-Hanssen, see Chap. 3). By 1921, Icelandic producers sent sorters, exporters and ship owners to Barcelona to understand the quality demands of the Catalonians.

To sum up, generic Norwegian products are sold to the highest-paying customers, independent of location and market. The Norwegian industry produces lower-valued, semi-processed products (Trondsen, Chapter 9). The current strategy has left Norwegian salted fish exporters as price takers in the Spanish market. However, the Norwegians receive a lower price for their products than do their competitors (Xie and Myrland 2010).

14.7.3 More Market Integration of Norwegian Firms

Based on the current configuration of the value chain, the broken lines between the production and market boxes in Fig. 13.1 cannot be filled, and thus Norwegian producers cannot overcome cultural barriers in the market (cf. Johanson and Vahlne 2003). For this reason, they will miss out on market-specific knowledge that could add value to their product. Internationalization is usually an incremental process in which a firm begins to export their products through an agent, then they might form a sales subsidiary and finally (in some cases) begin production in the host country (Johanson and Vahlne 1977, 2003). Norwegian processors have exported salted fish to the Spanish market for centuries without integrating their production to become insiders in the market. The majority of Norwegian firms still sell through an agent that is supported by the Norwegian Seafood Council's promotional activities and financed through a levy paid by all Norwegian seafood companies. The agent acts as an "ambassador" for the Norwegian seafood industry in a number of markets and provides Norwegian producers with general market information. In this way, they constitute a quasi-contact between the individual Norwegian producers and the Spanish market.

As already mentioned, the Norwegian–Spanish value chain for salted fish is market co-ordinated and fragmented, with little vertical integration between the production stations in the chain. As a result, only one of the two feedback loops proposed by Dicken (2007) is functioning optimally. The flow of money is working, but the flow of information is less successful. "Lack of knowledge about foreign markets and operations is the main obstacle to internationalization and knowledge can mainly be developed through experience from operations in those markets" (Johanson and Vahlne 2003, p. 89). Lindkvist (cf Chapter 7) shows how Norwegian production conventions and lack of knowledge about the Spanish market create barriers to significant Norwegian market participation. Therefore, psychic distance, which prevents the flow of information to and from the market, is a challenge for producers in the internationalization process (Johanson and Vahlne 1977, 2003).

Psychic or cultural distance must be overcome to make both feedback loops work. Differences in language, education, business practices, culture and industrial development create psychic distance, but it can be overcome by trust (Dulsrud 2001). Several Iberian production companies mentioned trust as the most important characteristic of their supplier. A Portuguese company completed contract negotiations with a major supermarket chain together with their Norwegian supplier and another Portuguese producer. Based on the strength of their relationship, they avoided becoming part of a captive, co-ordinated chain, which many producers have experienced in value chains driven by supermarkets. Trust is developed through relationships and mutual knowledge supported by institutions. Institutions reduce uncertainty and can secure predictable quality (Hayter 2004). A challenge in many seafood value chains, especially those based on wild captured fish, is to produce products with consistent quality from heterogeneous raw materials (see Chap. 13). Salted fish from Norway has been criticized for fluctuations in quality. What constitutes quality is continually adjusted according to shifting sets of criteria (Ponte and Gibbon 2005). Quality is constructed through the interrelationships among consumers, producers, traders, retailers and so on (Mansfield 2003). Quality can be good, bad, right, or wrong, and the market needs a number of different grades of quality and price categories. The perception of quality is culturally and socially embedded, and producers need insider knowledge to provide the level of quality that consumers want.

To overcome psychic barriers, the industry needs more market knowledge and experience. Firms must either be present in the market or need to find another mechanism by which they can secure and institutionalize the information feedback loop. In the next paragraph, we will suggest a cultural approach as such a mechanism.

14.7.4 A Cultural Knowledge Strategy

What we are looking for is a mechanism that can bring together the fragments in the chain and transfer the conventions and quality preferences from consumers to producers who are looking for added value. Here we will suggest a cultural knowledge strategy based on related variety, bringing together complementary pieces of market-relevant knowledge (Asheim et al. 2011, p. 894). Knowledge is more easily transferred from one sector to another when the sectors have complementary forms of competence (Asheim et al. 2011, p. 895). The cultural strategy should lead to the transfer of market knowledge and food culture from the regionally based consumer and market stations in Spain to the regionally based fishing and production environments in Norway. Several scholars (Lindkvist et al. 2008; Sánchez-Hernández 2011; Trondsen 2012) have pointed out the lack of insight about Spanish market conventions in the Norwegian salted fish industry. Norwegian producers have prioritized the collection of market information about factors that influence the prices of their standardized products, which is a market price orientation. To take value added advantage of Spanish consumers' product preferences, a deeper market knowledge and orientation is required. It is also essential that the right market knowledge be transferred according to business objectives. Asheim et al. (2011) refer to three relevant knowledge bases for market-oriented firms, analytical, synthetic and symbolic knowledge. Analytical knowledge (knowing why) is scientific, abstract knowledge, and it is often developed by research units. Synthetic or practical knowledge (knowing how) is problem solving, often through interactive learning with customers or suppliers. Symbolic knowledge (knowing who) is related to the creation of meaning and desire. The two first knowledge bases are present in the Norwegian salted fish industry, whereas the latter is more difficult to find with respect to understanding the consumer market culture. This lack of knowledge might influence the optimal working of the two first knowledge bases for the salted fish industry in Norway.

Salted fish is food with a strong tradition and culture, and a broad set of recipes. To spread symbolic knowledge and to ensure cross-fertilization across professional and sectoral borders, it is important to have informal meetings with end consumers and to hear the "buzz" of daily life. Contextual settings play a big role in symbolic knowledge (Manniche 2012). Symbolic knowledge is more often tacit and closely linked to specific socio-cultural contexts (Manniche 2012). In an interconnected chain, tacit information is exchanged through frequent face-to-face interactions (Gereffi et al. 2005). Thus, a relationally co-ordinated value chain, implemented through an internationalization process, is one way to transfer symbolic knowledge between market actors and producers at each end of the value chain. A problem will occur if the actor groups are uninterested in such interactions and knowledge exchanges. If the firms and producers in the north do not wish to be present in the market, maybe relevant representatives from the market could come to the northern producers?

In addition, when an industry uses a cultural strategy, the product itself plays a central role in building business interactions. In the value chain for salted fish, there are complementary sectors and complementary pieces of knowledge. The producers see the salted fish product as a commodity, whereas the consumers see it as their dinner. Asheim et al. (2011) explain that constructing a regional advantage and engaging in long-term development in a region depends upon the ability to apply current knowledge bases and competences to new sectors. The challenge is to create meeting platforms where knowledge transfers can take place. Tourism is a possibility, but tourist experiences in the food sector are normally unplanned, and the quality of the knowledge transferred is uncertain. A planned strategy would avoid the more uncertain aspects of incidental knowledge acquisition, as it would allow producers to structure meetings according to their explicit aims. In our case, chefs can function as representatives for symbolic knowledge.

In 2011, a bacalao festival was arranged in Honningsvåg in the North Cape municipality in Finnmark, northern Norway. Fisheries are important in Finnmark for employment and culture. A festival with chefs from important and exotic markets could increase the consciousness of people in production areas about cod as a food in market environments. This could occur by bringing the narratives and conventions from the markets to the producers to stimulate mutual learning. Variation between market and production districts would be exposed from district to district, hence transferring symbolic knowledge between the production environments.

In 2011, top chefs and musicians from Galicia in Spain were invited to Honningsvåg and the North Cape. This Spanish market could be a potentially important market for Norwegian salted fish. Galicia is a region with long fishing traditions. The market is relatively “small” and transparent, with about 2.8 million people. The consumption of fish is high, and bacalao salado is still on local menus. In the Bacalao festival, the Galician chefs prepared the salted fish, presenting multiple dishes, and they suggested possible new products to the industry. The chefs used the local vocational school as an arena for preparation and teaching. The locals were impressed by the various ways to prepare the fish. They were proud and astonished that a traditional product from their region was so valued in the Iberian Peninsula. The festival let market conventions, symbolic knowledge (in the form of narratives) and materialized heritage products (as food dishes) travel to the production part of the chain. There is justification in narratives and heritage products. The continuation of such a festival could be a way to renew even industrial (synthetic) and academic (abstract) knowledge platforms. An expansion of the festival could involve the exchange of cooking students between Norwegian and Spanish vocational schools. Such an exchange would be a dynamic and diachronic mechanism in the synchronic value chain structures. This strategy, together with the integration strategy, would be a mechanism for stimulating demand and production in the value chain structure. Norwegian and Spanish producers need to raise the consciousness of and interest in this traditional product among those in the coming generation to stimulate innovation by capitalizing on the traditions and knowledge platforms of the environments in the value chains that are dependent on each other.

14.8 Conclusion

Time is a central function in the development, change and continuation of deep value chain structures. This chapter began by introducing the synchronic and past-dependent deep structures in the Spanish salted fish market. Four deep structures or trends were proposed: the demand deep structure, which has had an astonishing stability in the markets for traditional products; the product structure, with change in the composition of products; the geographical pattern of conventions; and the value chain transaction pattern.

Based on these deep structures, we presented possible strategies for the Norwegian salted fish industry to sell their products, achieve long-term market positions and make a profit. The strategies must be understood as general advice for the industry. The first strategy was to continue the current regime of a generic product that is suitable for several markets in a market co-ordinated value chain. This strategy gives Norwegian producers independence and the flexibility to switch between markets. The current strategy is acceptable within the present institutional framework, which leaves the producers to implement variation in quality, but it may not be the most profitable strategy for the industry as a whole in the long run. The second strategy we proposed was more market oriented, advising producers to focus on internationalization to become insiders in the market and to participate in relational, co-ordinated value chains. Knowledge of both production and the market, and transferring knowledge from the one arena to the other is necessary to reach this goal. As shown by Sanchez-Hernández (see Chap. 4), Norwegian products have characteristics that the Spanish consumers appreciate, such as craftsmanship, sustainability, tradition, cleanliness, local foundation of products, and independence for fishers. Because of the lack of integration in the market, these domestic, civil and green conventions are not well communicated to the Spanish consumers, and Norwegian producers seem to be unaware of the value of these conventions.

The third strategy, a cultural knowledge strategy, can be seen as a continuation of real market integration based on heritage exchange and the learning of conventions from each other. This strategy thus tries to change the institutional framework that creates barriers to knowledge exchange and hinders integration in the value chain. The cultural knowledge strategy brings related knowledge together and makes visible the people and their food culture in the chain. The sector needs to renew its knowledge and to embed this knowledge where the products are produced as well as where they are consumed. The aim is to transfer knowledge from other places to the producers via food, cultural and educational institutions. The Norwegian side of this communication exchange may benefit by making it clear that Norwegians sell nothing but cod, free of injections other than brine with salt. This message could be communicated during face-to-face meetings and in festive parties but not through negative campaigns that could scare people away from consumption of salted fish in general.

The current quality standard sets criteria for participation in the Spanish market (see Chap. 6). Porter (1980) explains how new substitutes for a product can affect competition in an industry. In the case of the Spanish market, the introduction of phosphate has created an alternative to the traditional salted fish product. Norwegian producers cannot compete with this alternative product. Reduced Norwegian market shares in an increasingly differentiated product market for salted fish are related to (among other things) the use of phosphate. Although phosphate injected fish is illegal or unwanted from a health perspective, it obviously satisfies colour and taste conventions. There are other less preferred characteristics in the deep value chain structures, such as uneven quality, bad classification, low market knowledge and little integration, which function as market barriers. The Norwegian industry needs to understand that even though phosphate has changed the Spanish market, it is not the sole reason for their losses in this market. It may be that by focusing too much on the phosphate issue, the industry has missed other forms of action it could take.

Thus, even though phosphate is an important issue, it is not necessarily a complete explanation for losses in market share but rather an indicator of the industry's lack of market orientation. As long as the production conventions in the industry are focused on the quantity and price of traditional products, the Norwegian industry will most likely continue as before. Market-oriented conventions consist of attitudes, practices, preferences, and tacit knowledge that can only be acquired through interaction with the market, or representatives of the market.

In this chapter, we have tried to identify strategies that could increase the competitiveness of the Norwegian fishing industry and contribute to the creation of a more emancipatory policy for action. The strategies have certain contingencies and are based on initiatives that seem possible in the current Norwegian situation in which the government is focused on clean and unpolluted Norwegian seafood as the "world's best". The main contingencies are the formal and informal institutions. In a speech concerning the government's vision for the Norwegian seafood industry through 2030, the Norwegian Minister for Fisheries and Costal Affairs called attention to the need to improve quality in the catches and to develop quality products for the markets. She also said that Norway has the best raw material in the world and that it should be treated as such to provide quality products for consumers. Otherwise, Norway will lose (Berg-Hansen 2012). Even if bonds are forged between regions in Spain and regions in Norway, as suggested with the cultural knowledge strategy, and even if we have the institution of Seafood Norway, such bonds will not last if Norwegian producers do not make quality products. On the other hand, if Norway is to integrate into a market, it not only must build relationships in the form of a relational co-ordination of the value chain but also must create a friendly institutional environment (cf Xie, Chapter 11). The industry would improve by connecting the market and the production arenas. Internationalization is the recipe, but the industry cannot do it alone. The producers are part of a regulation regime that seems to narrow manoeuvring room and encourage production rather than market orientation. To take advantage of market-oriented value adding, the value chain actors need formal institutions to back them up. They need regulations that enable market-oriented flexibility in production. Norwegian producers would gain from a flexible institutional framework that enables more dedicated production and the possibility of producing both light salted and traditional products together with other desired products in Spanish and other markets.

References

Asheim BT, Boschma R, Cooke P (2011) Constructed regional advantage: platform policies based on related variety and differentiated knowledge bases. Reg Stud 45(7):893–904

Berg-Hansen L (2012) Regjernings visjon for sjømatnæringen 2013. http://www.regjeringen.no/nb/dep/fkd/aktuelt/taler_artikler/ministeren/taler-og-artikler-av-fiskeri-og-kystmin/2012/regjeringens-visjon-for-norsk-sjomatnari.html?id=698023. Accessed 2 Oct 2012

Chandler AD (1962) Strategy and structure: chapters in the history of the american industrial enterprise. M.I.T Press, Cambridge

Crevoisier O (2004) The innovative milieus approach: toward a territorialized understanding of the economy? Econ Geogr 80(4):67–379

Dicken P (2007) Global shift mapping the changing contours of the world economy. Sage, London

Dosi G (1997) Opportunities, incentives and the collective patterns of technological change. Econ J 107(44):530–1547

Dulsrud A (2001) Tillit og transaksjoner. En kvalitativ analyse av kontraktsrelasjoner i norsk hvitfiskeksport. Universitetet i Oslo, Avhandling for dr. polit-graden. Institutt for sosiologi og samfunnsgeografi

Gallart-Jornet L, Escriche I, Fito P (2005) La salaz′on de pescado, una tradici′on en la dieta mediterr′anea (Salted fish, a tradition within the Mediterranean diet). Editorial de la Universidad Polit′ecnica de Valencia. 2 ed. ISBN: 84-9705-918-2

Gereffi G, Humphrey J, Sturgeon T (2005) The governance of global value chains. Rev Int Polit Econ 12(1):78–104

Groff R (2000) The truth of the matter: Roy Bhaskars's critical realism and the concept of alethic truth. Philos Soc Sci 30(3):407–435

Hayter R (2004) Economic geography as dissenting institutionalism: the embeddedness, evolution and differentiation of regions. Geogr Ann 86B(2):95–115

Johanson J, Vahlne J-E (1977) The internationalization process of the firm—a model of knowledge development and increasing foreign market communities. J Int Bus Stud 8(1):23–32

Johanson J, Vahlne J-E (2003) Business relationship learning and commitment in the international process. J Int Entrep 1:83–101

Larsen HB, Lindkvist KB (2014) Generating generative mechanisms J Crit Realism 13(2):139–162

Lindkvist KB (2010) Mistrust and lack of market innovation, a case study of loss of competitiveness in a seafood industry. Eur Urban Reg Stud 17(1):31–43

Lindkvist KB, Sánchez JL (2008) Conventions and innovations: a comparison of two localized natural resource-based industries. Reg Stud 42(3):343–354

Lindkvist KB, Gallart-Jornet L, Stabell MC (2008) The restructuring of the Spanish salted fish market. Can Geogr 52(1):105–120

Manniche J (2012) Combinatorial knowledge dynamics: on the usefulness of the differentiated knowledge bases model. Eur Plan Stud 20(11):1823–1841

Mansfield B (2003) Spatializing globalization: a "geography of quality" in the seafood industry. Econ Geogr 79:1–16

Ponte S, Gibbon P (2005) Quality standards, conventions and the governance of global value chains. Econ Soc 34:1–31

Porter ME (1980) Competitive strategy. Free Press, New York

Sánchez-Hernández JL (2011) The food value chain as a locus for (Dis) agreement: conventions and qualities in the Spanish wine and Norwegian salted-cod industries. Geogr Ann Ser B, Hum Geogr 93(2):105–119

Sayer A (1992). Method in social science. A realist approach. Routledge, London

Sayer A (2006) Realism as a basis for knowing the world. In: Aitken S, Valentine G (eds) Approaches to human geography. Sage, London, pp 98–106

Sleipnes T, Johansen O (2012) Norske Konvensjonelle Produkter. Markedsplan 2010–2012. Revidert utgave 2012. Tromsø, Norges sjømatråd

Trondsen T (2012) Value chains, conventions and market performance: a comparative analysis of Norwegian and Icelandic cod exports to Spain. Can Geogr/Le Géographe canadien 56(4): 459–473

Whipp R (2001) Strategy: organizational. In: Smelser NJ, Baltes PB (eds) International encyclopedia of the social & behavioral sciences. Elsevier. http://www.elsevier.com/books/international-encyclopedia-of-social-and-behavioral-sciences/smelser/978-0-08-043076-8. Accsessed 22 April 2015

Xie J, Myrland Ø (2010) Modeling market structure of the Spanish salted fish market. Food Econ—Acta Agric Scand Sect C**7**:119–127

Yeung HW-c (1997) Critical realism and realist research in human geography: a method or a philosophy in search of a method? Prog Hum Geogr 21(1):51–74

Index

K. B. Lindkvist, T. Trondsen (eds.), *Nordic-Iberian Cod Value Chains,*
MARE Publication Series 8, DOI 10.1007/978-3-319-16405-2

FSC
www.fsc.org
MIX
Papier aus verantwortungsvollen Quellen
Paper from responsible sources
FSC® C105338

If you have any concerns about our products,
you can contact us on
ProductSafety@springernature.com

In case Publisher is established outside the EU,
the EU authorized representative is:
Springer Nature Customer Service Center GmbH
Europaplatz 3, 69115 Heidelberg, Germany

Printed by Libri Plureos GmbH
in Hamburg, Germany